The Wisdom of the Desert

James O. Hannay

IAP © 2009

Hannay, James O.

The Wisdom of the Desert / James O. Hannay – 1st ed.

   1. Religion

Cover Image

© Maxfx | Dreamstime.com

PREFACE

This little book is neither a critical examination of the earlier Egyptian monastic literature nor an historical account of the movement. It is nothing more than an attempt to appreciate the religious spirit of the first Christian monks. I do not know of any other similar attempt, though an exceedingly interesting study of the hermit life will be found in E. Lucius' *Das Menchische Leben des vierten und funften Jahrhunderts in der Beleuchtung seiner Vertreter und Gonnor.*

The collection of stories and sayings which I have translated, sometimes very freely, must be regarded merely as an anthology culled from the "meadows" of the literature of the desert life. There is much more which is worthy of a place in our devotional literature, and which, I hope, may in, the future be arranged and translated by men more fitted for the task than I am. I acknowledge gratefully the assistance I have received from two friends - Miss Bloxham and the Rev. C. S. Collins - whose sympathy with things that are high and holy has been a constant help to me in my work.

I have further to acknowledge the very great kindness of Father Andrew, S.D.C., who designed the drawings which both adorn this volume and interpret the spirit of the hermits' teaching.

After the MS. of this book was in the publishers' hands I received, through the kindness of Professor Zockler, of Greifswald, a copy of his recently published *Die Tugendlehre des Christentums* The work is of great importance for anyone engaged in the study of the ethics of monasticism, but I have not felt myself obliged to modify anything I have written. Professor Zockler's point of view and his object are entirely different from mine. He is scientific; I hope only to suggest devotional thought.

In the course of my Introduction I allude to the want of a critical study of the Apophthegmata. I am now informed by Dom E. C. Butler, O.S.B., that such a work is being prepared by Abbé Nau (Abbey Nau), and will soon be published in the *Patrologia Orientalis* by Firmin-Didot (Paris).

J. O. H.

Westport, Ireland, 1904

3

INTRODUCTION

Every kind of effort after good has found sympathy and help in Christianity. Nothing is more wonderful and nothing more suggestive of His divinity than the way in which the words and example of the Master have been found adaptable to the ideals which have possessed the souls of men in different ages and under various circumstances. There was a time when men were impelled to search for and express truth, the eternal truth of the nature and property of the Deity Himself. At that time the life of Christ presented itself primarily as a revelation. He set forth, under the conditions of time and space, the mysterious God whose seat is amid clouds and darkness, and yet who baffles human inquiry chiefly by the garment of impenetrable light in which He has decked Himself. In another age the religious spirit took a lower flight and allowed its activities to be dominated by a political conception. Whole generations spent themselves in the effort to realize upon earth a veritable kingdom of God. To these men Christ appeared as a monarch, whose will it was their ambition to realize perfectly. The people crowded below the altar steps, and the priests from above proclaimed, pointing the Lord to them, "Behold your King." He was, indeed, conceived of as very different from any earthly king. His crown was of thorns, His throne was a cross, His glory was humiliation. Yet it was essentially as a King that they conceived of Him. He was the Ruler of a visible kingdom, the Head of a hierarchy of governors, the promulgator of a polity and laws. For men of yet another generation religion found itself in the aspiration after personal liberty. Fear and ignorance had tyrannized over the earth - fear, the daughter of superstition; ignorance, superstition's handmaid. Minds which dared to question and doubt lived under a perpetual menace. Above all, the great tyrant was sin. Its fetters grew heavier on men's limbs, and checked the effort after progress. Then men came to think of Christ as a great liberator; their souls responded to the call, " Christ shall make you free." Since then the central point of religion has shifted again. In our time men no longer look to Christ to teach them truth. We have lost sight hopelessly of "the cloud-capped towers and gorgeous palaces" of the city of God upon earth. The naked individualism of the reformation period offers an inadequate view of life. We are inclined to doubt about the very existence of such a thing as liberty. We have discovered in Christianity a great incentive to philanthropy. Christ is for us, perhaps the man, perhaps the God, at least the One who fed men and healed them and taught them as none other ever did. Blindly sometimes, perplexedly always, we hurry to the hovels of the hungry and the bedsides of those who suffer even loathsomely; we build libraries and schools, being sure at least of this, that in doing these things we follow Him.

To all these various ideals Christ has been found entirely responsive. Each has found in Him a starting-point from which to escape the bondage of materialism. It has never, of course, been true that one great purpose has possessed the followers of Christ to the exclusion of every other. The conception of the gospel liberty lay quite consciously behind the enthusiasm for pure truth. The most faithful statesmen of the mediaeval Kingdom of God washed the sores of lepers and cast their cloaks over the shoulders of beggars on the wayside. The dominating conception of religion has always been permeated, leavened, tempered with conceptions of the Master's meaning which were strange to it. There has always been, besides, one great conception of religion which has existed along with each of the others in its turn. Christianity has always involved a hunger and thirst after righteousness. Always and everywhere Christians have felt the unquenchable desire to be good, and have seen in Christ the great example of perfection. There has been no age in the history of the Church in which the idea of imitating Christ has failed to make an appeal to the souls of the faithful.

Yet even this desire has had its period of special intensity, its peculiar region where it became for a while the expression of Christianity. During the fourth and fifth centuries, in, the deserts of Egypt and Palestine, the craving for perfection was more painful and more narrowly exclusive than ever elsewhere. Thousands of men and women, in response to a passionate hunger after righteousness, set themselves to become perfect, as the Father in heaven is perfect. They were not, indeed, careless about right belief and the holding fast of the faith. The accusation of heresy was a thing which seemed to them wholly intolerable. Yet to them the supreme importance of being good was so felt that it seemed of necessity to bring with it a true faith. "What is the faith? " asked a brother once. The abbot Pimenion replied to him, "It is to live always in charity and humility, and to do good to your neighbor." Their absorption in the pursuit of holiness made speculation seem vain and impious. "Oh, Antony," said the heavenly voice, "turn your attention to yourself. As for the judgments of God, it is not fitting that you should learn them." Nor must we think of the hermits as disregarding the claims which the Church made upon their obedience; still less as neglecting the claims of the poor and suffering. We shall see, later, how they thought about the Church, and how unjust it is to call them selfish. Here, first of all, it is necessary to understand that they were not chiefly theologians, or churchmen, or philanthropists, but imitators of Christ. Their desire was to be good. That they also believed rightly and did good followed - and these things, did follow - from their being good.

This aim of theirs ought not to be strange to us. Indeed, it cannot be. In the midst of our multiplied activities there is something in us which responds to the ideal of being, as well as doing, good. It is the WAY in which they sought to attain their end, and not the end itself, which is incomprehensible and generally repulsive to the modern mind. It is so, I think, mainly because it is so absolutely strange to us. Our imaginations refuse to aid us in the effort to realize a system of religious life based upon complete isolation from the world. To us the activities of life - the getting and spending, the learning and teaching, philanthropy, intercourse, and the opportunities for influence - constitute life itself. It is as difficult for us to form a definite conception of a life apart from the world, from business, society, and the movements of human thought, as it is to realize that life of disembodied waiting which we expect in Paradise. Yet this complete isolation was what the Egyptian hermits strove to attain; and if we are to appreciate the value of their teaching we must, first of all, grasp the fact that they were real men on whom the sun shone and the winds blew, men with local habitations, and not phantoms or unsubstantial figures in a dream. If we conceive a fourth-century traveler starting as Palladius did from Alexandria, we may suppose that he would journey due south, ad skirt at first the shores of what is now Lake Mariut. Along the barren and rocky margin of the lake, at spots as remote as possible from the track followed by caravans, he would find the hermitages of ascetics, who, like Dorotheus, maintained a comparatively close connection with the Alexandrian clergy. Leaving the lake and journeying still southwards over about forty miles of utterly desolate land, he would come to a long valley extending east and west between two ranges of mountains or table lands, covered with sandy flats, salt marshes, and dangerous rocks. This is the famous Nitrian desert. Here St. Amon built the first solitary cell. Here Evagrius Pontikus lived for about two years. Here Nathaniel was visited by the bishops. Here the "Long Brothers" lived, one of whom was the companion of St. Athanasius when he went to Italy. At the end of the fourth century the Nitrian mountains were dotted over with hermits' cells. The evenings were resonant with psalm-singing. On Saturdays and Sundays the brethren swarmed forth like bees for worship in their church. Five miles further south, still among the Nitrian mountains, lay a region so utterly desolate that it had not even a name, till the monks built over it and "christened" it *The Cells*. Further south still and towards the west lay the Scetic desert. It was a day's journey from *The Cells*. This is the most famous of all the monastic settlements. Its founder was St. Macarius the Great. We may reckon among the Scetic monks his two namesakes, St. Macarius of Alexandria and Macarius the Young. Here also, for the most part, dwelt Pior, Moses the Ethiopian, Paul the Simple, and the hermit Mark.* South-eastward, past Lake Arsino and Herakleopolis, lay St. Antony's birthplace, Coma. Here, no doubt, might have been seen the tombs into which he first shut himself, and across the river, the mountain on which he found his ruined fort. This mountain, which was called "the outer mountain," formed the home of smaller and less famous groups of ascetics. South-east from this, within a few miles of the Red Sea, lay "the outer mountain," to which St. Antony was guided by the heavenly voice. Perhaps this retreat was never shared with him by anyone except his chosen attendant and the few visitors who forced their way there in search of spiritual counsel. South from the "outer mountain," along the river, lay Oxyrynchus. This, even if we discount the figures of contemporary writers, must have been a great monastic city. In it monasticism took in organized ecclesiastical form. The church was served by priest-monks, and great communities of men and women carried on works of charity and evangelization. Still further south lay Lycopolis, the home of John the prophet. This man was celebrated as well for his wonderful obedience as for his spiritual gifts. Lycopolis may be reckoned the outpost of the monasticism of lauras and hermitages. Beyond it lay the organized monasteries of the disciples of St. Pachomius. During the lifetime of the founder of Tabennisi, nine monasteries carried out his rule. Of these the most famous was that which was ruled by Bgoul and afterwards by his nephew, Schnoudi. On the sea-coast, east of Alexandria, lay the settlements visited by Cassian. The Tannitic mouth of the Nile flows into what is now Lake Menzaleh. In Cassian's time this whole region was a desolate salt swamp. The sea flowed over it when the north wind blew, destroying all hope of fertility. On the hills, which came to look like islands, stood the ruins of villages forsaken by their inhabitants. It was a land

"Sea saturate as with wine."

Among the ruins and amid the surrounding desolation dwelt the monks who were the heroes of Cassian's earlier Conferences. No scene has seemed to me to convey more vividly at once the pathos and the nobility of the monk's renunciation of the world than this one. In Nitria and Scete the ascetic is at least remote from all remembrances of

6

common life. On the islands of Menzaleh he kneels in solitary prayer within the very walls where women once laughed to see their children sport. He gazes over brine-soaked swamps, which once were harvest-fields thronged with reapers. Westward from Menzaleh lay Lake Burlus. Between it and the sea stretched a desolate spit of sandy land, given up by farmers as hopelessly barren. This was the Diolcos described in the Institutes, and the eighteenth Conference. Here Archebius and his fellow hermits struggled for life in their inhospitable home, husbanding even their water as no miser would husband the most precious wine.

Thus we have five distinct and widely separated regions in which Egyptian monasticism existed and flourished during the fourth century. First, Nitria, with its offshoot *The Cells*; second, Scete; third, the region in Upper Egypt which came under St. Antony's more immediate influence; fourth, Southern Egypt; fifth, the sea-coast of the Nile Delta. In very close connection with these, so as to be predominatingly Egyptian in the tone of their monasticism, were the hermitages and lauras of south-western Palestine and the settlements in the Sinai peninsula. Outlying from the greater centers were single hermitages and small lauras, wherever the monks hoped to find solitude.

In many places life was supported only with extreme difficulty. Sometimes water had to be obtained by collecting and storing the dew which fell at certain seasons. Sometimes it was carried with immense toil from distant wells. There were districts where the hermits lived in constant dread of the irruption of barbarian tribes, which destroyed tranquility and even threatened life itself. Bands of wandering robbers sometimes rifled the cells of their miserable furniture, or captured, insulted, and injured the hermits. At other times the silence of these retreats became so awful, that the hermit was startled into uncontrollable emotion by the chance shout of some shepherd-boy who had driven his goats too far; or came to find the rustling of dry reeds in the wind an almost insupportable noise.

For the most part in the deserts north of the Thebaid the monks saw very little of each other. Even the inhabitants of grouped cells led almost solitary lives. On Saturdays and Sundays they met for public worship and perhaps a common meal, but during the rest of the week they lived alone in their cells, or with a single disciple. If the monk were wise, he worked. Sometimes he wove mats or baskets. These were afterwards exchanged by the hermit himself or his disciple for the necessities of life in some neighboring village. If the cell lay too remote from human habitation to permit of such traffic, the mats or baskets were accumulated in piles, and in the end burnt. They had fulfilled their function, and were got rid of that way as well as in the markets; for the hermit was not a tradesman. He worked, not for wages, but lest the devil might tempt him in his idle hours. Sometimes a garden was cultivated around the cell. The hermit struggled with drought and barrenness until he produced a little stock of vegetables. Sometimes his cell was happily placed where date palms grew. He watched his fruit against the depredations of wild birds. Nothing is more striking than the insistence of the greater hermits on the necessity for labor of some sort. It was from their experience and their illuminated introspection that St. Benedict learnt the truth on which he built a great part of his rule - "Idleness is the enemy of the soul."

Besides working, the monks prayed. Hours every day were spent in prayer, which must have been more of the nature of meditation than intercession. In the intervals of prayer and work they sang or said psalms, and often repeated aloud long passages from the prophets. Books were scarce among them, and we read of monks visiting. each other for the purpose of learning off by heart fresh passages of Holy Scripture. The attainment of unbroken monotony was a thing greatly to be desired. Perfect quietness was the monk's opportunity for spiritual communion with God. Therefore they regarded restlessness and the wish for change as a sin to be fought against. Long periods of unbroken monotony were liable to produce in the monk a spirit of irritable peevishness and discontent with his surroundings, which was recognized as subversive of true spirituality. They called this state of mind "accidie" and held that it was the work of a special demon. The monk felt its force chiefly during the long hours of daylight when he grew weary of praying and shrank from the petty tasks which had to be performed around and within his cell. The spirit which tempted him to accidie was "the demon which walketh at noonday." It was chiefly in order to conquer this sin that the monks worked as hard as they did at even quite useless tasks. They knew that it was fatal to try to avoid the attacks of accidie by seeking change of scene and fresh interests. Their one hope lay in labor and remaining quietly in their own cells.

Sometimes the monotony of life was broken for the monk by the arrival of a stranger. The more famous among them were so frequently visited, that the quiet which was necessary for their own religious life was seriously interfered with. St. Antony, for instance, was obliged to retire to his remote "inner mountain" in order to avoid his numerous visitors; and Arsenius made it a rule during one period of his life to receive no visitors under any pretext whatever. For most of the monks, however, the arrival of a stranger was a comparatively rare occurrence. Sometimes, if his cell lay between two great settlements, he would be called upon to entertain brethren who were traveling from one to the other. If he lived within reach of any town, clergy and pious laity came occasionally to his cell as to a kind of retreat, looking for spiritual refreshment from his words, and participation in his prayers. Aspirants after the glories of the monastic life visited hermits, of whom they happened to have heard, in search of advice. On all such occasions it was the duty of the hermit to entertain his visitors. Hospitality was as much a duty in the Egyptian deserts in the fourth century as in the mediaeval monasteries of the Benedictines. The monk brought out his little store of dainties and spread a "feast" for his guests. Here is the account of a "sumptuous repast" offered to a traveler. "He set before us salt and three olives each, after which he produced a basket containing parched vetches, from which we each took five grains. Then we had two prunes and a fig a piece. When we had finished our repast, he said to us, 'Now, let me hear your question.'" The hermit not only afforded his guest the best food at his command, but, in a true spirit of hospitality, he ate with him. Very often this necessitated breaking a fast which he was keeping, or departing from his ordinary rule of life. Sometimes, for the sake of his guests, he even omitted portions of his evening prayers, or said them secretly after his visitors had gone to sleep; for the duty of hospitality came before almost every other.

Sometimes the monks themselves deliberately broke the monotony of their lives, and went on an expedition to visit some renowned saint. They did so to seek advice for the conquering of some besetting sin, or to inquire the meaning of a passage of Holy

Scripture over which they had long meditated in vain. Often they asked vaguely for "a word," so they called it, from the saint; that is, for any exhortation that might be offered, any fruit of a religious experience deeper than their own. These answers, or "words," were eagerly treasured in the memories of those who heard them. They passed from mouth to mouth as opportunities for intercourse occurred. The brethren in a laura were eager to hear from a returning monk what he had learned on his visit. Thus we read of the brethren in the Scetic desert crowding round St. Macarius on his return from the "inner mountain," and plying him with so many questions that he was interrupted in his account of what St. Antony had said to him. Naturally collections of specially striking sayings and anecdotes came to be made in the various lauras. I imagine that quite early in the fourth century the monks took a pride in remembering as many as possible of the "words" which they had heard. Soon collections of them began to be written down, and probably before the end of the fourth century there existed in the greater lauras written lists of famous sayings. These local collections embodied stories from all sources, and very frequently the names of the original authors are altogether lost. In the course of the fifth century larger collections came to be made, probably by travelers who either had the opportunity of inspecting local collections or heard the stories from old monks. If we believe that the collection given by Rosweyd in Book III. of his Vitae Patrum was actually made by Rufinus himself, we have one dating from the end of the fourth century. In these larger collections the stories are arranged in one of two ways, either they are grouped under the names of their authors, where these are known, or in chapters according to the subjects they deal with. Thus, in the great Greek collection, (published in Migne P.G. LXV.) all the anecdotes bearing the name of St. Antony are grouped together, and those with the name of Besarion together, and so on. In the collections of which Rosweyd published Latin translations, all the stories illustrating, for instance, such virtues as humility and patience come together, without regard to the names of their authors. That these various collections were made independently of each other, and from different sources, is seen in the fact that anecdotes which are quoted as anonymous in one collection bear the name of an author in another. Sometimes the same saying is attributed to different authors, and sometimes what is substantially the same story appears in several different forms. Thus there is a fine saying attributed in one place to Sisois in the form - "Qui peregrinatio nostra est, ut teneat homo os suum," which appears twice elsewhere as anonymous in the shorter form "Peregrinatio est tacere." It seems likely in this case that the longer form is the nearest to the words originally used. I have endeavored to give the sense of this saying - translation I take to be impossible - in Chapter 14, number iii.

It is from the collections of these "words of the fathers," which have been published by Rosweyd and Migne, that the greater part of the translations in this volume are made. That they are genuine remains of the teaching of the early monks of the Egyptian and South Palestinian deserts I have no doubt whatever. At the same time, it is only fair to warn the reader that these collections have never been critically edited, and that other collections exist which have not yet been published. It is much to be desired that some competent scholar would undertake the labor of editing those which exist only in MS. and critically examining the whole mass of this literature.

In order to appreciate fully the marvelous spiritual beauty of their teaching, it is necessary for the modern reader, in the first place, to realize that the hermits were actual living men, and to make an effort to understand the kind of lives they lived. It is as a help to such effort that I offer the first part of this introduction. In the second place, the reader must try to clear his mind of certain prejudices which exist against the hermits and their way of life. It is to the consideration of these prejudices that I have given up the following portion of this introduction.

When the "sumptuous" repast, which I have just described, was finished, the abbot Serenus said to his guests, "Let us hear your question." One of them replied, "We want to know what is the origin of the great variety of hostile powers opposed to men and the difference between them." In reply, the monk discussed for several hours the nature of principalities and powers, of Beelzebub, of the Prince of Tyre mentioned by Ezekiel, of Lucifer, and of the crowds of evil spirits which hover in the atmosphere around us. Such questions and such discussions inevitably raise in our minds a prejudice against the men who engaged in them. We leap at once to the belief that there must have been in their minds a tendency to fantastic and entirely barren speculation. I am not inclined to either minimize or explain away the fact that the whole literature of early Egyptian monasticism is shot through and through with evidences of a belief in the reality, personality, and power of demons. The monks believed that every temptation which came to them was the work of a special demon. There was the demon of anger, who provoked brethren to quarrel with one another; there was the demon of despair, whose voice reminded the penitent of former sins, and urged the impossibility of his salvation; there was the demon who walked at noonday - he lured the monk into the sin of accidie; there were demons of gluttony, of pride, of vainglory, of covetousness. The demons had the power of assuming appalling or seducing forms, of becoming visible and palpable. Monks heard them clamoring and roaring, felt their blows, smelt them when they were present. Victorious fiends who had terrified their victims into submission or lured them into sin vanished amid peals of derisive laughter. Defeated, they departed with lamentable and awe-inspiring shrieks. Men who had experienced the ferocity and insistence of these powers of evil cannot be accused of being unpractical or merely speculative when they discuss their nature. To the Egyptian monk the power of devils was, except only the power of God, the most practical and pressing question which could be discussed.

Yet, even if we grant this, our prejudice remains. The whole apparatus of these powers of evil is strange and incredible to us. Good and evil as tendencies or opposing principles we understand, or think we understand. We smile at what seems the rude anthropomorphism which sees a demon personally present in the natural cravings of a starved body, and hears a voice through the broken sleep of a long series of solitary nights. We dismiss such tales as no doubt meant to be true, but in reality only the delusions resulting from prolonged fasts and the morbid phenomena of hysterical enthusiasm. It would be possible, of course, to urge, in defense of the hermits' beliefs, that the apostles thought substantially as they did about the powers of evil. We might parallel even such stories as that of the beating of St. Antony from the book of the Acts of the Apostles. It might be urged that our Lord's own teaching forces us to believe in just such personal, audible, and palpable spirits of evil as the hermits say they strove against. Unfortunately such appeals to authority, even to the supreme authority of all, are of comparatively little use to us. They may result in an irritated assent to the conclusions of a syllogism, or check the utterance of words of contemptuous incredulity; they can neither compel our sympathy nor silence the protests of our imagination. It seems better, if we wish to get into spiritual touch with the hermits, to approach these demon stories in another way. We must be conscious that we have never hungered and thirsted after righteousness with such intensity as these early monks did. We have not been driven, as they were, into a divine madness by the unsatisfied desire for perfection. Until we have felt as they did, struggled as they did, forced our way into the region of spiritual effort in which they lived, have we any right to feel sure that our interpretation of their experiences is the true one? It may be, too, that we allow ourselves to be prejudiced against the hermits' version of what they endured by the bald simplicity with which the tales are told. St. Athanasius' doctrine, so far as the reality and, personality of the powers of evil are concerned, is in no way different from that of St. Antony. It is because he philosophizes in the light of history, instead of narrating experiences, that his doctrine does not shock us. We are not irritated by the conception to which the poet Milton has given utterance in his Ode.

Peor and Baalim

Forsake their temples dim,

With that twice-battered God of Palestine;

And moond Ashtaroth,

Heaven's queen and mother both,

Now sits not girt with tapers' holy shine;

The Lybic Hammon shrinks his horn,

In vain the Tyrian maids their wounded Thammuz mourn.

They feel from Juda's land

The dreaded infant's hand;

The rays of Bethlehem blind their dusky eyn

Nor all the gods beside

Longer dare abide,

Nor Typhon huge ending in snaky twine:

Our Babe, to show His Godhead true,

11

Can in His swaddling bands control the damnd crew.

Milton's demons are in no way essentially different from those which attacked the hermits in the deserts. Yet, because his conception is expressed in gorgeous words and sonorous rhyme, our imaginations do not refuse to rise to it. Neither the speculations of the great father nor the language of the poet are any argument for the reality of the demons they describe; but the fact that we can enter sympathetically into their thought does seem to suggest that it is not the substance, but the manner of the hermits' demon stories, which revolts us. It is, after all, quite in accordance with the spirit of the apostolic age to conceive of the ancient gods as demons, whom Christ had driven from the images where they lurked and the temples in which they were worshipped. It requires but a simple application of the Lord's words to enable us to think of these malevolent beings trooping in mortified disgust to desert places, there to wander, seeking in vain for rest. It was along some such line that the thoughts of the hermits moved. St. Antony and the others went into the wilderness with the belief that they were entering upon a region still the property of demons, as the whole world had been before the coming of the Lord, In their journeyings along the reaches of the Nile they stumbled upon the ruins of once gigantic temples. Huge images frowned upon them, painted figures, "delicate and desirable," smiled to allure them. Amid the vast monotony of the desert, where man's insignificance is impressed upon him, nothing seemed strange because it was supernatural. The monks conceived themselves as fighting a final Armageddon with the already broken forces of the Prince of this world; or, when the ascetic conception of St. Paul appealed to them, as "filling up that which was lacking in the sufferings of Christ," and consummating the final expulsion of that kind which goeth not out but by prayer and fasting. Along such lines of thought it is perhaps impossible for our minds to move with a sense of comfortable security. Yet our imagination ought not to be wholly incapable of making such an effort to appreciate their view of life as will enable us to understand their teaching and sympathize with their effort.

Another prejudice against the hermits and their teaching arises from our extreme dislike of their severe physical asceticism. We are disgusted by the details of their war against the flesh, and we rise in revolt against their ideal of crucifying their bodies. In our time the popular conscience has come to have an almost morbid dread of pain. Perhaps the fact that our religion is largely dominated by the idea of philanthropy is simply one expression of a widespread shrinking from the suffering which is the common lot of humanity. We have almost ceased to speak of this shrinking as cowardice in the case of the individual who dreads pain for himself. We frankly stigmatize as brutal the infliction of pain as punishment, which former generations regarded as an edifying spectacle. It is therefore peculiarly difficult for us to appreciate the position of men who deliberately refused to gratify the cravings of their bodies, who joyfully sought out suffering for themselves, and did not hesitate to encourage others to "crucify" their bodies. It is not to be denied that our position with regard to physical asceticism finds a specious justification. We may ask whether it is believable that the Creator can be pleased with creatures who reject His gifts and nullify the instincts which He implanted in them; whether we can imagine the tender and compassionate Savior demanding as the price of following Him such renunciation as St. Antony's. Such questions are not easy to answer. They open up the whole problem of the place of asceticism - asksis, exercise, discipline -

in the Christian life. It is not possible here to enter on such a discussion. There are, however, two considerations which, if they in no way solve the general difficulty, yet may serve to mitigate the prejudice which the special austerities of the Egyptian hermits arouse in us. In the first place, we must remember that these men aimed at perfection, and hoped to attain it by a literal imitation of Christ. Now Christ on one occasion fasted forty days and forty nights. It is quite natural that men who aimed at imitating Him should fast and should try to make their fasts like His in their severity. Christ also lived a virgin life. It is only to be expected that His imitators should determine to be virgin too - virgin in body and, if we may use the expression, virgin in mind, according to His explanation of the meaning of purity. Christ describes Himself as homeless and poor. He was worse off than the foxes and the birds. We cannot wonder that the desire of imitating Him has led men to renounce their property and to accept homelessness as one of the conditions of living perfectly. Christ's life terminated in the torture of the cross. To "crucify" themselves - the word is a favorite one with them - was part of the ideal of the monks. They meant by "crucifixion" every kind of hardship, privation, and pain home voluntarily for Christ's sake. It was nothing else than an attempt at participation in the sufferings of Christ. Once, on a feast day, a disciple moistened his master's bread with a few drops of oil. The old hermit burst into tears, and said, "My lord is crucified, and shall I eat oil?" Christ proclaimed that a man could not be His disciple without hating his father and mother and his own life also. The words came as a challenge to those who wished to follow Him, a challenge which the monks accepted literally. Of course, it is possible to say that all such simple acceptance of Christ's teaching and literal following out of His sayings is a narrowing, even a perverting, of the spirit of the gospel, and that it leads to a kind of life quite different from that which the Lord contemplated for His disciples. This may be so. To discuss it is to enter upon that larger question of the place of asceticism in the Christian life which we have already passed by. Whether we are prepared to recognize the monastic ideal as the ultimate and loftiest conception of the teaching of Christ or not does not for our present purpose seem to matter. The hermits' life was certainly an attempt to imitate Christ and obey His commandments. No one who loves the Lord can refuse to sympathize with men who, even mistakenly, have tried very hard to follow Him. The second consideration which I wish to urge in mitigation of our prejudice against the extremity of the hermits' physical asceticism is this. They never regarded it as anything but a discipline, a means to an end. They have been accused of being the slaves of a mechanical theory of virtue, of imagining that religion consisted in outward observances, of teaching that fasting and watching were righteousness. There is hardly any accusation possible which would be more decisively disproved by an appeal to the facts of the case. That it should have been made and repeated, as it has been, is a very curious instance of the confidence with which we are all inclined to dogmatize about things of which we are almost ignorant. Probably never, except in the age of the apostles, has the purely spiritual aim of all religion been kept more steadily in view than it was by the hermits. The best of them - and it is only from its best men that the true spirit of a movement can be learned - never for one single instant let slip the truth that no practice or discipline is of any use at all except in so far as it helps towards the attainment of the perfection which is in Christ Jesus. No one will be inclined to deny that it is possible to pick out of the literature stories of excesses which seem to us monstrous. There were many among the hermits who never rose above the idea that asceticism was an end in itself. But the excesses were discouraged and the mistaken idea condemned by the leaders of the movement. Fasting, virginity, labor, the reading and recitation of Holy Scripture, vigils, meditation, and even prayer itself, were looked upon simply as ways of arriving at a perfect life. There is no need to discuss whether or not they mistook the way. Even supposing that they did, at least the end they had in view was one which we must recognize as very great. It is possible, in spite of the evidence of accumulated Christian experience, that a man is hindered, and not helped on the road which leads to union with God, by fasting and watching and poverty, yet since this union is a thing which we also seek, we should, at least, approach with sympathy the study of the teaching of men who made for the goal by a way which was neither broad nor easy.

One more prejudice remains to be noticed, and this is one which has most to do with alienating our sympathy from the early monks. It has been said - there is no comment on monasticism which we hear more frequently - that the hermit life was a selfish one, and therefore essentially remote from the spirit of Christ. There is a very obvious retort to this accusation which, in spite of its obviousness, is not so superficial as it seems. The charge is directed against men who gave up everything that is usually counted as desirable. Renunciation like that of the hermits is not usually a symptom of selfishness. It comes from the lips of a generation who have found the service of Christ not incompatible with the full enjoyment of all life's comforts and most of life's pleasures. Perhaps, however, this retort, like most others of its kind, misses the real true point of the charge. The hermits are called selfish because they aimed at being good and not at being useful. The charge derives its real force from the fact that philanthropy, that is, usefulness to humanity, is our chief conception of what religion is. We appeal to the fact that Christ went about doing good, and we hold that the true imitation of Him consists in doing as He did rather than in being as He was. The hermits thought differently. Philanthropy was, in their view, an incidental result, as it were, a by-product of the religious spirit. Here, no doubt, there is a great gulf fixed between us and them. There is a difference of ideal. It is possible to aim at doing good, and snatch now and then, as opportunity offers, a space for the culture and of spirituality, for the "making" of the soul. It is possible also to shape life for the attainment of perfection, welcoming, as it may happen to offer itself, the chance of usefulness. The latter was the ideal of the hermits. Is the former ours? Surely the purest altruism will decline to accept it. We recognize, when we are at our best, that what we ought to aim at is that good should get done, and not that we ourselves should do it. The faithful soul, even when most pitiful of suffering, will still desire less to be useful than to be used in the cause of humanity. Impatience, that glorious impatience to be up and doing which we cannot but admire, rebels against delay and indirect approach. The evil around us is so clamorous for amendment that it seems like a betrayal to spend our strength any way but in the combat with it. Yet it remains, at least for the student of history, a question whether in the end, there is not more good accomplished for humanity through the agency of those who, in the first instance, only aim at being good. The case of the Egyptian hermits is an illustration of what I mean. They did not aim at doing good. This is why we call them selfish. Yet certainly there was accomplished through them a great work for religion and for the Church. We can only guess at how great an incentive to piety their lives, viewed from far off, were for Christians, who remained "in the world." We know that many men, clergy and laity alike, visited the hermits, sought and, we cannot doubt it, received from them fresh spiritual strength, rekindled in the desert cells lamps that had gone out for want of oil. We can only guess, too, at what their share was

14

in the great battle for the catholic faith. How much did St. Athanasius owe to them when he stood against the world? It was no small thing for him to know that there stood behind him men whom no court party had any bribes to buy, whom no emperor's frown had any power to terrify. The student of their literature will remember also that they did something for the material benefit of the Egyptian people. I do not insist upon the cures they wrought, or the devils they cast out of those possessed. Some of these stories belong to the region of the miraculous, though others are, and more no doubt will be, recognized as natural by the scientific mind. Apart altogether from these miracles, the hermits did an immense, but now quite unrecognized, amount of charitable work. Many of them earned a great deal more than they needed to spend, and all that they could spare was given to the poor. They appointed some of their number to oversee the distribution of their alms. They not only fed the hungry and relieved the destitute with whom they came into actual contact, but they sent camel and boat-loads of food to the poor in the great Egyptian cities. They tried to alleviate the misery of the prisoners confined in gaols. On at least one occasion they organized a collection and distribution of food on a large scale in a famine-stricken district. We shall, surely, not want to quarrel with a way of life which in fact proves to be very useful, even according to our own standard of usefulness, because in the first instance it aimed at something else. It is not however only, or even mainly, by their work for their own generation that the usefulness of the Egyptian hermits must be judged. They were the spiritual fathers of the monks of the west. It was to the Egyptian fathers that all the great founders of western monasticism looked back. St. Martin of Tours, Cassian, Benedict of Nursia and his later namesake of Anian, all drew their inspiration from the lives of St. Antony and his followers. The work which the western monks did for mediaeval Europe is written large across the pages of history. It is recognized even by writers who are out of sympathy with the monastic ideal. It is not necessary to describe the beautiful monastic charities for which our poor - laws have proved but a dismal substitute. We are ready to grant that the mediaeval monasteries were useful in their day. Ought not their usefulness to be reckoned for righteousness to the Egyptian hermits, who were the fathers of all monasticism? The Benedictine Rule, the parent of all the great rules down to the time of the Mendicant Orders, was nothing but the systematic adaptation of the teaching and experience of the Egyptian hermits to the needs of western life. So long as the western monks, under any rule, remained true to the old ideal of trying to be good in simple imitation of Jesus Christ, they also did good and were, as we say, useful. It is only when they forget or turn away from this ideal, when are touched with the spirit of the world, or set themselves to the accomplishment of some policy, that their organizations tend to do mischief. From this point of view the usefulness of the hermits far outlasted their own generation. Through them was effected a great good which could not have been foreseen. It is perhaps just because they denied themselves the satisfaction of aiming at usefulness that they were so greatly used. This seems to be one of the laws of the divine government of things. The Lord Himself suggests it when He says: "Seek ye first the kingdom of God and His righteousness, and all these things shall be added unto you."

It seems quite possible then that what is called selfishness in the hermits, may be in reality the loftiest altruism. If so, the gulf between their ideal and ours is not so great that the heart cannot cross it. It is only needful that we should see clearer and think deeper than we do, that we should be less sure that only we have grasped the meaning of the Master's

15

life. It is in the hope that the study of them may make for clearer vision, deeper thought, and most desirable humility that I offer these fragments of the wisdom of the desert to those who sincerely desire to be the friends of Jesus Christ.

* Since writing the description of the relative positions of Nitria, Cellia, and Scete, I have read the very valuable note (No. 14) in Dom Cuthbert Butler's *Lausiac History* (No. 2). He suggests a solution of the they geographical problem so different from mine that I think it right to give his words.

"Though the three authorities (Palladius, Cassian, and Rufinus) differ in their figures, they still agree as to the fact that Scete was distant from Nitria a long journey across the desert; and as they had all three visited Nitria, and as Palladius and Cassian claim to have actually made the journey between Nitria and Scete, their evidence as to the main fact must be accepted. The danger of losing one's way on the journey is illustrated by Palladius' story of a monk who died of thirst while travelling from Scete to Nitria or Cellia. . . . Now if Scete lay a day's journey to the *south* of the Wady Natron, it is difficult to understand how there can have been easy communication between it and Terenouthis; yet many passages show that such was the case (see AmŽlineau, GŽographie, 493); e.g. when the Mazices made an irruption into Scete it was to Terenouthis that the monks fled; but if Scete was several miles to south of Nitria, it would have been much more natural them to have gone on the line of the present track towards Cairo."

Dom Butler then cites a passage from Ptolemy, and adds: "Ptolemy thus places the Scetic region to the north of Nitria. If he is correct, and I am disposed to believe he is, Scete was that portion of the Libyan desert which between the Delta and the Wady Natron, some fifty miles across. And if that be so, Cellia was situated in this desert, six or seven miles north of Nitria; while still further to the north or north-west, in the heart of the Scetic desert, lay the monastic settlement of Scete."

The greatest weight must be attached to anything which so competent an authority as Dom Butler says on the subject of monastic Egypt, and I ought, perhaps, to give up at once the idea that Scete and Cellia lay to the south of Nitria. One great objection, however, to the northern site still weighs with me. Scete appears always to be regarded as more difficult of access than Nitria. If it lay to the north, would it not be easier to reach from Alexandria and even form a stage on the journey from that city to Nitria?

CHAPTER 1

The Hidden Treasure

The Kingdom of Heaven is like unto treasure hid in a field; the which when a man hath found, he hideth, and for joy thereof goeth and selleth all that he hath, and buyeth that field.

- *St. Matt.*, xiii. 44.

He that findeth Jesus findeth a good treasure, yea, a Good above all good.

- *The Imitation of Christ*, ii. 8.

THE desire of knowing the Lord is often the first impulse of the religious life. The human soul repeats the request of the Greeks who came to Philip, "Sir, we would see Jesus." The desire - it is only in such paradoxes that religious experience seems able to express itself - already witnesses to the fact that the soul, which feels it, has seen Jesus. The contradiction may he stated nakedly without the fear of making discord in the Christian consciousness. He who has seen Jesus once, still sees Him, and therefore must desire to see Him.

As the clearness of the vision varies, so does the intensity of the desire. For some men it is dim, as the reflection in a mirror that is cracked and tarnished. The shape is uncertain and its lines wavering. Jesus is not so seen as to shut out the possibility of seeing the other things which present themselves within, behind, or beside the mirror which reflects Him. Men who behold Hun thus, desire to see Him in a very real way; yet their desire does not become the master passion of their lives, so that all other desires grow faint or are excluded altogether. They wish for other things as well as the vision of Jesus. Certainly they will only wish for those other things whose possession is not inconsistent with the seeing of Him. Yet they do wish for other things. For a few men the mirror is more perfect and the reflection much clearer. To them the vision is so desirable that having once seen Jesus they thenceforth see nothing but Jesus, hope and strive for nothing else of any sort except to see Jesus.

The difference between these two ways of seeing is a difference in vocation. We cannot explain it. We no more seek either to explain it or to alter it than we seek to explain or alter the fact that it was St. John and no other who saw "in the midst of the seven candlesticks one like unto the Son of Man, clothed with a garment down to the foot, and girt about the paps with a golden girdle." We simply recognize that they who see clearest are they who have the higher vocation, and we know that their way in life must be the harder way, as our human nature reckons hardness. It is they who when they have found the treasure hid in the field are so absorbed with the desire of possessing it that they sell all they have; that is to say, they, having seen Jesus, see nothing else in life at all desirable except Jesus. It must always seem to most men a strange and very hard thing to give up everything for the sake of a spiritual gain. To those who have seen the vision and heard the vocation it is, save for the weakness of the flesh, not a hard thing. It is written in the gospel of them that "for joy thereof" they go and sell all that they have.

Among those who have had the vocation none have had it more certainly than the hermits of the Egyptian deserts. They, because they had seen very clearly, and in their daily lives continued to see very clearly, were, of all men, most absorbed in the desire of seeing Jesus. The stories and sayings in this chapter show us the intensity and the meaning of their desire. To the hermit Macedonius the pursuit of his God was a toil like a hunter's, but inexpressibly more absorbing. He cannot think of ceasing from his hunting. The abbot ,John knew that no enticements could seduce away his soul from entering upon the fruition of Jesus. The abbot Arsenius, who had been once a courtier in the Emperor's palace, recognized that all other desires must give way before the desire of seeing Jesus. Very wonderful is the perception of the abbot Allois of the remoteness of the soul which dwells with God from all else except God.

The lives of the hermits supply us with a very perfect example of the paradox of religious experience. They, if ever any men, grasped the true value of the hidden treasure. They, more than most men, realized the condition of obtaining it, and sold all that they might purchase it. Yet their lives afford an almost terrible example of the intensity of the continuous struggle which the purchasing entails. They saw Jesus, therefore they desired to see Him, and their desire forced them to pursue that holiness "without which no man shall see the Lord." This is the keynote of their lives. They were determined, at all cost, in some way to become good; because having seen Him very clearly, they knew that they would not get to see Him unless by His grace they grew to be like Him.

I

How the hermit Macedonius witnessed that it is not strange to do for the sake of possessing the Lord what men do willingly for smaller gains.

A certain captain of soldiers, who took a great delight in hunting, once came in search of wild animals to the desolate mountain where Macedonius dwelt. He was prepared for hunting, having brought with him men and dogs. As he went over the mountain he saw, far off a man. Being surprised that anyone should be in a place so desolate, he asked who it might be. One told him that it was the hermit Macedonius. The captain, who was a pious man, leaped from his horse and ran to meet the hermit. When he came to him he asked, "What are you doing in such a barren place as this is?" The hermit in his turn asked, "And you? What have you come here to do?" The captain answered him, "I have come to hunt." Then said Macedonius, "I also am a huntsman. I am hunting for my God. I yearn to capture Him. My desire is to enjoy Him. I shall not cease from this my hunting."

II

A word of St. Basil to one who was unwilling to sell all that he had in order to buy the field wherein the treasure is.

A certain Syncletius, a senator, renounced the world. He divided his property among the poor, but kept back some of it for his own use. To him St. Basil said, "Truly you have spoiled a senator, but you have not made a monk."

III

A word of the abbot Arsenius, him who left the emperor's court for the desert, seeking God; and resigned his wealth that he might take the hidden treasure.

"If we seek God, He will appear to us. If we hold Him fast, He will remain with us."

IV

The word of one who knew how good a thing it is to know of nothing in the world, but to know of Jesus.

The abbot Allois said "Except a man say in his heart, 'I and God are alone in the world,' he will not find peace."

V

How the enticements of the world have no power to lure back again the soul that has once possessed Jesus.

The abbot John said: "There was an exceedingly beautiful woman who dwelt in a certain city, and she had a multitude of lovers. A great man, one of the nobles of the city, came to her and said, 'Promise that you will be mine and I will wed you.' She gladly promised, and being his wife went to dwell with him in his palace. Afterwards her other lovers came seeking her and found her not. When they heard that she had become the nobleman's wife, they said one to another, 'If we go up to the door of the palace, it will be plain that we are seeking her, then, without doubt, we shall be punished. Let us go to the back of the house and whistle to her, as we used to do when she was free. When she hears our whistling she will certainly come down to us.' They did as they had planned, and the woman heard their whistling. Hating greatly even to hear them, she went into the inner parts of the house and shut the door upon herself. Now this woman is the soul of a man. Her husband, the nobleman, is Christ. The palace is the eternal mansion of the heavens. They who whistle for her are the demons."

VI

A comparison of one who desires to attain the eternal treasure to an archer who turns his eyes away from everything except his mark.

A man will despise all things present as being transitory when he has securely fixed the gaze of his mind on those things which are immovable and eternal. Already he enjoys, in contemplation, the blessedness of his future life. It is as when one desires to strike some mighty prize - the prize is virtue - which is far off on high, and seems but a small mark to shoot at. The archer strains his eyesight while he aims at it, for he knows how great are the glory and rewards which await his hitting it. He turns his eyes away from everything, and will not look save thither where the reward is placed. He knows that he would surely lose the prize if his strained sight were turned away from the mark even a very little.

VII

How a man cannot possess the heavenly treasure and at the same time cling to the pleasures of earth.

The abbot Arsenius was once asked by the abbot Mark why he fled from the society of men. He replied, "God knows it is not that I hate men. I love them well. But I cannot dwell both with God and with men. There are multitudes of heavenly beings and many virtues, but all their wills are one, and they come of one will. Among men it is otherwise. Their wills are many, and they pull us different ways. I am in this strait. I cannot leave God, for that is how I think of it, to dwell with men."

CHAPTER 2

On Being Crucified with Christ

If any man will come after Me, let him take up his cross, and follow Me.

- *St. Matt.*, xvi 24.

He who enters upon the way of life in fear bears the cross patiently. He who advances in hope bears the cross readily. He who is perfected in charity embraces the cross ardently.

- *St. Bernard, Sermon I. on St. Andrew's Day.*

I have received the cross. I have received it from Thy hand. I will bear it, and bear it even unto death, as Thou hast laid it upon me.

- *The Imitation of Christ*, iii. 36.

The agony of Christ will last till the end of the world; we must not slumber during this agony.

- *Pascal.*

ALL vision of Jesus includes some vision of His cross. It is not possible to think of Him without in the end arriving at a contemplation of his cross. The soul which loves to dwell upon the tender figure of the Good Shepherd will not be able to escape from the stern truth - "He giveth His life for the sheep." He who meditates faithfully upon the nativity will find that the shadow of the cross falls even over Bethlehem. In its effort to appreciate the King in His beauty, the mind is thrust back from the contemplation of "Him that liveth" to the recollection of the other word - "And was dead." The general consciousness of Christian people, of all ages and all races, has steadily recognized that the one thing needful is to know Christ crucified.

It is possible to think of the cross of Christ simply as the symbol of the great suffering borne for us. There is an endless depth of meaning in such an apprehension of the cross. The man, whose mind is so enlarged that he can contemplate the mystery of the life given that he might live, is penetrating into the divine love. He may go from what is bright into regions of yet intenser brightness, until he stops blinded by his nearness to the very God Himself. It is possible also that at some point of its meditation the soul may catch the infection of the love of God, and come to burn with a reciprocal love for Him who so loved men. Then the vision of the cross and suffering of Christ begets a sympathetic desire to suffer with Him. This desire in itself is so natural that we may regard it as one of the instincts of our nature. It comes within the experience of every one who has ever felt a great love for parent, brother, wife, child, or friend. When the object of our love suffers, we desire and even try to suffer too. We feel ourselves outraged at the thought of claiming a passing pleasure while one who is very dear to us lies in intense bodily pain. It is not that we expect our refusal of enjoyment to in any way assuage his sufferings. It is simply that we cannot pursue our own pleasures at such a time.

Now the sufferings of Christ had become very vivid and real to the minds of the hermits. Just as we instinctively shrink from laughter when one who is very dear to us lies dying, so they, because they loved Him greatly, desired to deny themselves pleasure and even to accept the burden of pain. This is the meaning of the aged Palaemon's refusal to eat food dressed with the unaccustomed luxury of oil. This seems also to be the meaning of the repeated use the hermits made of the word "crucifixion." Their fastings and vigils, their endurance of heat and toil, were spoken of as "crucifixions," because they conceived that in these sufferings, voluntarily borne, they were taking their part in the sufferings of Christ upon the cross. They even spoke of the diseases and physical evils which came upon them, independently of their own wills, as "crucifixions," for they knew that pain which is unavoidable may be so borne as to render it in reality a taking up of the cross.

Of course, there was always present to their minds the commoner thought that self-crucifixion was of benefit to the soul. They felt, as we are hidden to feel, that "our way to eternal joy is to suffer here with Christ. They analyzed the good that comes of suffering and deliberately courted it as a means of drawing near to God. Yet always at the back of such reasomlings there lay the feeling that suffering was borne with Christ, as well as for the sake of attaining His eternal joy. This idea of sympathetic fellowship in suffering is what gives its peculiar beauty to the hermits' interpretation of the words, "If any man will come after me, let him take up his cross and follow me." It is this thought which, while it adds a pathos to the stories of their lives, certainly helped them in winning that grace of perseverance which would not be satisfied until it reached "the goal of being crucified with Christ."

I

Of what it means to take up the cross with Christ.

Perhaps some man will say, "how can a man carry his cross? How can a man who is alive be crucified? Hear, briefly, how this thing may be. The fear of the Lord is our cross. As, then, one who is crucified no longer has the power of moving or turning his limbs in any direction as he pleases, so we ought to fix our wishes and desires, not in accordance with what is pleasant and delightful to us now, but in accordance with the law of the Lord in whatsoever direction it constrain us. Also, he who is fastened to a cross no longer considers things present, nor thinks about his likings, nor is perplexed with anxiety or care for the morrow, minor is inflamed by any pride, or strife, or rivalry, grieves not at present insults, nor remembers past ones. While he is still breathing in the body, he is dead to all earthly things, and sends his heart on to that place to which he doubts not he shall shortly come. So we, when we are crucified by the fear of the Lord, ought to be dead to all these things. We die not only to carnal vices, but to all earthly things, even to those indifferent. We fix our minds there whither we hope at every moment we are to go.

II

Of one who feared because God took the cross he bore from him.

There was a certain old man who was frequently sick and feeble. One whole year it happened that no sickness of any kind troubled him. He wept on that account, and was sorely afflicted, saying, "Thou hast left me, O Lord, and art unwilling to come to me this year."

III

Of the hermit Palaemon, how he desired to crucify his body because the Lord was crucified.

When the holy time of Easter came Palaemon said to his disciple St. Pachomius, "Prepare some special food for us to-day, since this is a feast day for all Christians throughout the whole world." Then St. Pachomnius, prompt ever in obedience, did as the old man bade hum. After their prayers were finished Palaemon went to the table to eat. When he saw there oil added to the usual food he burst into tears and smote his hands against his forehead, saying, "My Lord has been crucified, and I - shall I eat oil?"

IV

How the desire of being crucified with Christ will keep a man in the narrow way though he see others departing from it.

A certain elder was once asked, how a monk can avoid being offended and disheartened, when he sees others giving up the hermit life and returning to the world. He replied - "Watch the dogs which hunt hares. One of them only, perhaps, sees the hare and chases it. The others see nothing but the dog in full chase, so they run with him for a while and then grow weary and give up. The one that sees the hare goes on chasing it until he catches it. He takes no heed of the steep hills, nor of the thickets, nor of the brambles in his way. Sometimes his feet are flayed and pricked with thorns, yet he does not rest until he catches it. So it is with the monk who seeks Christ and gazes steadfastly on the cross. He takes no notice of the things which vex and offend him. He cares for nothing but attaining the goal of being crucified with Christ."

V

Of the narrow way which leadeth unto life.

A certain elder was once asked, "What is this which we read - 'Strait and narrow is the way?'" The old man replied, "The narrow way is that on which a man does violence to his own imaginations, and cuts himself off from the fulfilment of his own will. This is the meaning of that which was written of the apostles, 'Behold we have left all, and followed Thee.'"

CHAPTER 3

Being Dead to the World

The Lord - when the Jews spit on Him and buffeted Him and smote Him with their hands, when Peter denied Him thrice, when the priests and elders accused Him, when the soldiers mocked Him and scourged Him - answered not. He neither rebuked them, nor defended Himself, nor reviled again, nor cursed those that persecuted Him.

My son, in many things it is thy duty to be ignorant and to esteem thyself as one dead upon the earth, and to whom the whole world is crucified

- *The Imitation of Christ*, iii. 44.

Thou oughtest to be so dead to such affections of beloved friends, that (so far as thou art concerned) thou wouldest choose to be without all human sympathy.

- *The Imitation of Christ*, iii. 42.

THE idea of being dead to the world is very closely connected with that of being 'crucified with Christ.' Indeed, we may say that dying to the world is one of the ways in which a man accomplishes the crucifixion of himself with Christ. The hermits pressed the metaphor of being dead to its last possible conclusions. This they did with that simple and childlike directness which is one of the great charms of all their teaching. St. Macarius, for instance, in no way sought to evade the force of the metaphor. To him to be dead meant simply to be, so far as the world is concerned, a dead body. The body that is buried neither resents insult nor warms to praise. The perfect Christian must be equally deaf to blame or flattery. Anub's acted parable enforces the same thought, as does St. Antony's treatment of the stone outside his cell. Dorotheus, whose story Palladius tells, grasped at the ideal of deadness from a different side. To him it was the insistent claim of his body for consideration rather than the mental passions of anger or pride which kept him conscious that the world still claimed him as its citizen. He aimed at silencing his body's cries for rest and ease. In each case, whether the warfare is waged against the mind or the flesh, the ultimate aim of it is to reach a state in which the claim of God is everything, and the ways in which the world strives to distract from it nothing at all.

There is something in the tone of the stories which illustrate this ideal of deadness to the world that recalls the Stoic teaching about apathy. Indeed, so like is their moral to the maxims of Epictetus that we cannot wonder at the later monks having adopted his Encheiridion as a book of devotion. Yet between the ideal of the hermits and that of Epictetus there is a very real difference. The Stoic taught that a man should be dead to blame or praise. He praised an askesis which, as it were, insulted bodily desire. The strength by which a man attained this splendid apathy was pride. It was because the thief or the slanderer had no real power to hurt, that the philosopher was in a position to be indifferent to their injuries. To the Stoic the wrongs done to him were not to be resented, because when properly considered they were not serious wrongs at all. They had no power to affect the inner man - the soul - the only part of him that mattered. The view of the Christian hermit was entirely different. He made no attempt to persuade himself that injuries and wrongs were anything else than real injuries and wrongs. His soul stood in no proud isolation from their influence. To him neither praise nor blame were, or ought to be, matters of indifference. The one was a danger to be shrunk from, lest his soul should suffer; the other was a possible stepping-stone to the perfection which is in Christ Jesus. Thus, if St. Macarius' teaching about the dead bodies who were praised and blamed seems to be nothing more than the doctrine of Epictetus, we see that this was only part of all that the saint meant when we read of his holding back from the companionship of those who praised him, and seeking eagerly the society of his traducers. If Anub seems to teach nothing but the Stoic apathy in his acted parable with the stone image, we realise how much further the general teaching of the hermits went when we read Zacharias' comparison of a monk to a garment trampled into the dust. In truth, the hermit did not strive, like the Stoic, to be himself sublimely indifferent to all except his higher self, but rather strove to lose himself altogether since self, in his view, was of the world, and to find a new self in God.

I

How Zacharias, the disciple of the abbot Moses, showed that the followers of the Lord must accept such treatment as the Master received.

Certain brethren once came to the abbot Moses, and asked him to speak to them some word of exhortation. He turned to his disciple Zacharias and urged him, saying, "Do you speak somewhat to these brethren." Then Zacharias took off his cloak, and, laying it on the ground, trampled on it. "Behold" he said, "unless a man is thus trampled on he cannot be a monk."

II

The Abbot Sisois finds the secret of peace in the imitation of the suffe7ings of Christ.

The abbot Sisois said, "Suffer yourself to be despised. Cast your own will behind your back. Stand free from the cares of the world. Then you will have peace."

III

The parable which the abbot Anub acted, meaning to teach thereby that the disciple of Jesus must be dead alike to inmlt and to praise.

Once a tribe of Mazici burst into the Scetic desert, and killed many of the fathers who dwelt there. Seven of the fathers found safety in flight, among whom were the abbot Pimenius, and another older abbot called Anub. These seven came in their flight to Terenuthi. There they found an ancient temple of some heathen god, now deserted by the worshippers. Into it they entered, meaning to dwell together for a week without speaking to each other, while each sought a place where to build his solitary cell, for in the Scetic desert these seven had lived as hermits.

Now, there was in the temple an image of the ancient idol. The, abbot Anub guessed the thought of dwelling together which had entered the minds of the brethren. He therefore, when he rose in the morning, used to cast a stone at. the face of the idol. In the evening he used to speak to it, and say, "I have done wrong. Pardon me." On the Sabbath day, when the brethren met together, the abbot Pimenius said to him, "How is it that you, a Christian man, have for a whole week been saying to an idol, 'Pardon me?'" The abbot Anub replied to him, "I did this for your sakes. When I cast stones at the idol, was it angry? Did it speak to rebuke me? When I asked pardon of it, was it pleased? Did it boast?" The abbot Pimenius answered, "Surely no, my brother." Then said the abbot Anub, "We seven are here together. If we wish to remain thus and yet find profit for our souls, this idol must be our example. When one of us is insulted or vexed by another, he must not get angry. When one of us is asked for pardon by his brother, he must not be puffed up. If we are not willing thus to live together it is better for each of us to depart to whatever place he wishes." Then all of them fell upon, their faces to the earth, and promised that they would do as he advised.

IV

Dorotheus the Theban, being persuaded that the flesh and the spirit are contrary one to the other, mortified the flesh with his exceeding toil. This he did that he might be partaker of the life which is in Jesus.

All day long, even in the heat of summer, Dorotheus used to collect great stones along the shore of the sea. Though now an old man, he never ceased from the labour of building cells of the stones which he gathered. These cells he gave to hermits who could not build for themselves. Once a certain man asked him, "Why, my father, do you in your old age persist in slaying your body with such toil as this in the intolerable heat?" He answered, saying, "My body is slaying me. I am determined therefore to slay it."

V

How St. Macarius taught the meaning of the apostle's words "Dead with Christ," "Buried with Christ."

A brother once came to the abbot Macarius and said to him, "Master, speak some word of exhortation to me, that, obeying it, I may be saved." St. Macarius answered him, "Go to the tombs and attack the dead with insults." The brother wondered at the word. Nevertheless he went, as he was bidden, and cast stones at the tombs, railing upon the dead. Then returning, he told what he had done. Macarius asked him, "Did the dead notice what you did?" And he replied, "They did not notice me." "Go, then, again," said Macarius, "and this time praise them." The brother, wondering yet more, went and praised the dead, calling them just men, apostles, saints. Returning, he told what he had done, saying, "I have praised the dead." Macarius asked him, "Did they reply to you?" And he said, "They did not reply to me." Then said Macarius, "You know what insults you have heaped on them and with what praises you have flattered them, and yet they never spoke to you. If you desire salvation, you must be like these dead. You must think nothing of the wrongs men do to you, nor of the praises they offer you. Be like the dead. Thus you may be saved."

VI

Of bearing with evil men, and how a man may thus be a peacemaker since he will refuse the occasion of strife.

A certain hermit saw some men toilsomely bearing a dead body to the burial, and said to them, "You do well that you thus bear the dead. You would do better still to bear with the living. Then you would be makers of peace, and inherit the blessing of the Lord."

VII

Of two things by which a man is hindered from being truly dead to the world.

The abbot Pimenius said, "That monk may truly reckon himself dead to the world who has learnt to hate two things, ease for his body, and the vainglory which cometh of the praise of men."

VIII

St. Antony teaches that a monk should be like a rock.

St. Antony spoke to the abbot Ammon saying, "You have still a long way to advance in the fear of the Lord." Then leading him forth of the cell he showed him a rock and said to him, "Go, hurt that rock. Beat it unmercifully." This he did, and St. Antony asked him whether the rock made any answer. He said "No." Then St. Antony said to him, "You must attain to the position of the rock and not know when anyone is trying to hurt you."

IX

How the abbot Macarius used to avoid the conversation of those who honoured him, and preferred to talk with men who offered him insults.

When anyone came respectfully to the abbot Macarius, desiring to hear some exhortation from him, he received no answer at all. But if anyone came despising Macarius and did violence to him in such words as these, "Lo you there, father Macarius! You used to be a camel-driver, and steal the natron[1]. How your master used to beat you when he caught

---

[1] (*n.*) Native sodium carbonate.

28

you robbing him!" willingly, even joyfully, Macarius used to speak to such a man of whatever he wished to hear.

CHAPTER 4

How We Ought to Return Good for Evil

Love your enemies, bless them that curse you, do good to them that hate you, and pray for them which despitefully use you.

- *St. Matt.* v. 44.

"My friend," said the bishop, "before you go take your candlesticks." He went to the mantlepiece, fetched the two candlesticks, and handed them to Jean Valjean.

"Now," said the bishop, "go in peace, Jean Valjean, my brother, you no longer belong to evil, but to good. I have bought your soul from you."

- *Victor Hugo, Les Miserables.*

We can pass quite naturally from the consideration of what the hermits taught about dying to the world to the stories which illustrate the ideal of returning good for evil. Indeed, when we think of death to the world, as evidenced by the patient endurance of wrong, the passing over to the thought of returning good for such evil is so gradual that it is hard sometimes to decide under which heading to place some particular story or exhortation. Yet there seems to be a real difference between the two groups of stories. Those which illustrate death to the world are concerned chiefly with the inner life of the hermit himself. Their interest centers in the condition of his soul and its approach to the ideal of suffering with Christ. In those which deal with returning good for evil the point of moral stress shifts, and the action of the hermit is thought of mainly as it affects the man who did the injury. This distinction is not merely arbitrary. It goes back to the twofold way in which the cross of Christ is viewed. When the contemplative soul dwells mainly on the sufferings of the cross, and love is aroused to attempt to take a sympathetic share in the pain, we have the stories of self-crucifixion and death to the world. Where, on the other hand, the thought of the death on the cross as a sacrifice is predominant - that is, of a life given that others might have life - we come upon a series of stories in which the main stress lies upon the effect on others of our imitating Christ. This thought comes very clearly before us in the beautiful interpretation which the abbot Poemen gives of the words: "Greater love hath no man than this, that a man lay down his life for his friend." Here the listener's mind was taken back at once to the cross of Christ and he is shown, not so much how he is to be partaker of the sufferings as how he is to share in the Lord's work of sacrifice. The same thought is present, if less obviously, in the story of the old man who kissed the hands of the brother who had robbed him.

There is nothing in this part of the hermit's teaching which should be strange to any Christian. It is impossible, in the face of the Lord's words in the Sermon on the Mount, to accuse their conduct even of exaggeration. All that we find wonderful is the extreme simplicity with which they understood the sayings of the Lord and adopted them as a practical rule of life. For most men there is need of certain explanations, of an effort of the intellect, of casuistry, before the Lord's commands can be reconciled with the maxims which direct the ordinary life. It is necessary to write some gloss beside. the saying - "If any man take away thy coat, let him have thy cloak also." Otherwise we cannot but be conscious of a divergence between the conduct which life seems to render necessary and that which is recommended by the Lord. For the hermits and their admirers no such necessity existed. They took the commands of Christ and obeyed them as if such obedience involved no absurdity. Strange as it must seem to many men, their literal obedience resulted not in an impossible deadlock and the dissolution of social relationships, but in an incomparably great type of character, and frequently in the reclamation of sinners whom methods less apparently absurd would have confirmed in their viciousness. Thus the conduct of Anastasius towards the brother who stole his book was, from the world's point of view, absurd. It is impossible to conceive of the continued existence of any society in which the majority of men not only refused to punish, but actually rewarded thieves. In the assistance which St. Macarius rendered to the robbers who were rifling his cell, the absurdity reaches a climax. Yet Anastasius rescued his brother's soul from perdition, and if in the case of St. Macarius we know nothing of the effect of his acts on the robbers, our hearts are at least enriched with the conception of a man whose spirit was the very spirit of the Lord.

It is perhaps especially interesting to notice that even in the case of postulants, whose hearts shrank back from the prospect of offering the other cheek to the smiter, there is no effort to evade the direct literalness with which the hermits interpreted our Lord's commands. They hoped, apparently, to be somehow excused from obedience. It did not occur to them to cast round for an explanation of the words which would enable them to think of themselves as obeying while they refused to obey literally. The view which the hermits took of what the world calls justifiable resistance of evil is well exemplified in the story of the brother to whom the abbot Poemen wrote a letter. The hermit's action when the robbers attacked him seems to have been most natural and right. He called for help, and the robbers were caught and imprisoned. Yet on account of what he had done, his conscience would not let him rest. Poemen explained to him where his sin lay and why his conscience troubled him. He ought, so it seems, to have acted as St. Macarius did under such circumstances. But, it may be asked, what then would become of society, of the security which civilization has brought with it for property and life? I do not suppose that the hermits could have answered the question. I can imagine only that they would have parried it with another. What otherwise is to become of the commandments of Christ?

I

How an old man blessed one who injured him.

A certain brother came to the cell of an elder, one well known among the brethren for his holiness. Entering in, he stole the food which was there. The old man saw him, but did not accuse him. He only labored more diligently to supply again what he had lost, saying in his heart, "I am sure that my brother must have been in great need, for else he would

31

not have stolen." In spite of his toil, the old man came to endure great suffering for want of food. At last he was brought even to the point of death. The brethren, knowing only that he was dying, came and stood round his bed. Among them he saw the brother who had stolen his food. "Come hither to me," he said to him. Then taking his hands and kissing them, he said to those who stood around, "I pay my thanks to these hands, brethren, for because of them I am going, as I trust, to enter the kingdom of heaven."

Then that brother was stricken to the heart, and repented. He also in the end became an eager monk, wrought upon by the deeds of the elder which he saw.

II

How the abbot Sisois taught a brother that the desire of vengeance separates a man from God.

There was a certain brother who had suffered an injury at the hands of another. Coming to the abbot Sisois, he explained the wrong which he had suffered, and then said, "My father, I desire to be avenged." The old man begged him to leave his avenging in the hands of God, but he persisted, saying, "I cannot rest until I have well avenged myself." Then Sisois said to him, "Since your mind is altogether made up with regard to this matter, I need not reason with you. Let us, however, pray together." Thus saying, he arose and began to pray in these words: "O God, Thou art no longer needful to us. We do not require Thy care of us. We ourselves are willing, yea, and are able to avenge ourselves." As soon as the brother, who had desired vengeance, heard these words, he fell at the old mass feet and begged for pardon. "As for him with whom I was angry," he said, "I shall not in any way contend with him."

III

A doctrine concerning injuries done to us by which we may escape from the danger of being angry, and even turn such wrongs into a source of profit for our souls.

A certain brother, who had been injured by another, came and told the story of what had happened to one of the elders. This is the reply which the elder made to him: "Set your mind at rest concerning the wrong done to you. The harm was not meant for you, but for your sins. In every temptation to anger or hatred that comes to you through the act of man, accuse not him who does the injury. Say simply, 'It is on account of my own sins that this, and things like this, happen unto me.'"

IV

Of the one which may be reckoned supreme amongst the commandments of the Lord, both inasmuch as it is beyond all difficult to be kept, and ako in that the keeping of it makes us fellow-sufferers with Him.

A certain brother came to an elder seeking some word of exhortation. "Tell me," he said, "of some one commandment, such that I may keep it, and thereby attain unto salvation." The old man answered him, "When men do wrong to you and revile you, endure and be silent. To do this is a very great thing. This is above all other commandments."

V

The abbot Poemen teaches that they who have grace to keep this commandment are very sharers in the death of the Lord upon the cross.

A certain brother once questioned the abbot Poeman, saying, "What is this word which the Lord says in the gospel, 'Greater love hath no man than this, that a man lay down his life for his friend?' How may one do such a thing?" The old man answered him, "Perhaps a man may hear from his friend some word which insults and angers him. Perhaps it is in his power to speak back to his friend in like manner. If then he chooses to endure in silence ~ if he does violence to himself, being fully determined to speak no angry word, nor any word to hurt or vex the other ~ then, verily, this man lays down, in sacrifice, his life for his friend."

VI

The dealings of St. Antony with certain brethren who wished to be perfect, but sought for some other way than the way which the Lord taught.

Certain brethren once came to Saint Antony and besought him to speak to them some word through which they might attain unto the perfection of salvation. He, however, said to them, "Ye have heard the Scriptures. The words which have come from the lips of Christ for your learning are sufficient for you." when they still pressed him, begging that he would deign to speak some word to them, he said, "It is taught in the gospel that if a man smite you on the one cheek you are to turn to him the other also." They then confessed that they were not able to do this. St. Antony answered, "Is this too hard for you? Are you willing to let such a man strike you on the same cheek twice?" They said, "We are not willing," hoping to be told of some easier thing. But he said to them, If this, too, is beyond you, at least do not render evil for evil." Again they answered him as they had done before. Then St. Antony turned to his disciple who stood by, and said, "Prepare some food and give it to these men, for they are weak." But to the brethren who had inquired of him, he said, "If you cannot do one thing and will not do another, why do you come seeking a word of exhortation from me? To me it seems that what you need most is to pray. By prayer perhaps you may be healed of your infirmity."

VII

A story of St. Macarius, showing how he would not resist one who robbed him.

The abbot Macarius, when he dwelt in Egypt, once had occasion to leave his cell for a little while. At his return he found a robber stealing whatever was in the cell. St.

33

Macarius stood and watched him, as one who was a stranger might watch having no interest in what was stolen. Then he loaded the robber's horse for him and led it forth saying, "We brought nothing into this world. The Lord gave and the Lord hath taken away. According to his will so things happen. Blessed be the name of the Lord."

VIII

How the abbot Anastasius would not resist an evil done to him, and thereby won his brother's soul.

Anastasius had a manuscript written on vellum which was worth a great sum of money, for it contained the whole of the Old and New Testaments. It happened that a certain brother who came to visit him, seeing this manuscript in his cell, coveted it. At his departure he stole it. After a little while Anastasius desired to read something in his manuscript. He searched for it but could not find it. Then he understood that this brother had stolen it. He was unwilling, however, to send after the thief or to ask him to restore the property lest, perhaps, he might add a lie to the sin of his theft. The brother who had committed the theft went straightway to a neighboring town in order that he might sell the manuscript. When one came to buy it, he named a certain price. Then the buyer said, "Let me have the manuscript that I may find out whether it is worth so much." Receiving it, he went straightway to the abbot Anastasius, and said to him, "My father, I pray you look at this book, and tell me if it is worth such a price. It is for such a sum that a certain man seeks to sell it to me." The abbot Anastasius answered him, "It is a good book, and is well worth what you are asked for it." Then he who was about to buy returned to the seller, and said, "Take the price you name. I have showed the book to the abbot Anastasius, and he told me that it was a good book, and well worth your price." Then the seller, he who had stolen it, asked, "Did the abbot Anastasius say anything more to you about it?" The other said, "No. I have told you all he said." Then the thief replied to him, "I have thought again about the matter, and I am not willing to sell the book at all." This he said, being cut to the heart. He hastened to the cell of the abbot Anastasius, threw himself upon the ground, and with tears of penitence besought the abbot that he would take back the book. But Anastasius refused, saying, "Go! and my peace go with you, brother. Take the book for your own. I give it freely to you." But he persisted weeping and praying, and he said, Unless you take back the book, father, my soul will never anywhere find peace." At length he took back his own book. Afterwards that brother remained with the blessed Anastasius, sharing his cell with him until the day of his death.

IX

How, by meeting evil which was done to him, a certain nwnk was led on to do a deed which grieved him greatly.

There was a certain great hermit who dwelt in the mountain called Athlibeus. It happened that he was attacked by robbers. He at once cried out, and the brethren who dwelt in the neighboring cells ran to his assistance and captured the robbers. They were sent to the nearest city, and the judge condemned them to be put in prison. Then all those brethren were sad because on their account the robbers had been put in prison. They went to the abbot Poemen and told him all that had happened. He wrote a letter to the hermit, whom the robbers had attacked, in these words: "You have betrayed the robbers to punishment. Remember that was not your first act of betrayal. First you betrayed yourself. Unless you had been betrayed by the evil within into resisting the wrong done to you, you would not have made that second betrayal of which you now repent."

X

How the injuries done to us by evil men are means whereby we may attain perfection.

There was once a monk who observed this rule of life. The more anyone injured or insulted him, the more eagerly he sought that man's company. This he did because, as he was wont to say, "Those whose company I seek are they who afford me the opportunity of perfection. They who speak well of us and bless us set our paths about with stumbling-blocks. It is they who deceive us."

CHAPTER 5

On Charity to Sinners

Whoso shall cause one of these little ones which believe on Me to stumble, it is profitable for him that a great millstone should be hanged about his neck, and that he should be sunk in the depth of the sea.

- *St. Matt.* xviii. 6 (R.V.).

Deal not roughly with him that is tempted; but give him comfort, as thou wouldest wish to be done to thyself.

- *The Imitation of Christ,* i. 13.

THE teaching of the hermits about charity to sinners is more closely connected than appears at first sight with their view of the obligation to return good for evil. To the indifferent and lukewarm Christian the only sin which is really hard to forgive is a sin against himself. Every man finds it difficult to forgive one who has injured or slandered him; but for the lukewarm Christian it is much more difficult than for him who has really tried to live after the pattern of Christ. On the other hand, the man who has little or no real enthusiasm for righteousness finds it easy to forgive a sin which is unlikely to result in injury to himself. For him who is fired with a real love for goodness it is just as hard, and sometimes perhaps even harder, to forgive a sin which is only a sin against God as one which directly affects his own well-being. This is the reason why men who are in earnest about religion are so often, and, we must add, so justly, accused of hardness and want of charity. Very often the worldly man's charity is nothing but indifference, is a fatal weakness masquerading as a virtue. Very often the good man's hardness and bitterness are the defects which naturally result from an earnestness which is very admirable.

It will not be denied that the hermits were genuine enthusiasts for righteousness. We are not, therefore, surprised when we find that their enthusiasm frequently led them into harsh and stern judgments of sinners. What does astonish us is that the greatest of them rose superior to the defect which seems almost the inevitable concomitant of their virtue, and displayed a large-heartedness and a charity like Christ's. In this respect, as in every other, the name of St. Antony stands pre-eminent. We realize how great a man he was when, remembering his own steadfastness, we read his parable about the ship which was almost wrecked. Even more striking is the word of Besarion when the priest cast the sinning brother out of the church. Such tenderness to sinners can have only sprung from a deep and real love. The men who had it certainly realized the meaning of the words "He that dwelleth in love dwelleth in God, for God is love." The way in which a man, whose zeal for righteousness had extinguished his charity, came to realize his fault is very wonderful. It is very easy for us to understand the character of this hermit who pronounced unhesitatingly the doom of expulsion on the brother who had sinned. We can picture him - stern, upright, uncompromising, a rigorist. The type - at least, in the pages of history - is only too familiar. What is hard for us to realize is the swift spiritual perception of the man when he grasped the meaning of Pastor's enigmatic parable and the ready self-abasement of his confession. Perhaps even more unexpected than the charity to those who have sinned is the large-heartedness of the same abbot Pastor in his answer to the brother who wished to make a feast. Anub has his code of right and wrong, clear-cut and inflexible. In Pastor we see a great humanness. His words are like the Lord's when He said of the woman, "She hath done what she could."

We enter upon difficult ground when we touch upon the story of the monk who told a lie to save his brother's soul. The polemical casuist will find in such a tale a text for disputation and hair-splitting. To the elders among the brethren this aspect of the story does not seem to have occurred. They glorified, not indeed the lie, but the love of the brother who lied, and honored him as one who imitated Christ even to the point of laying down his life for his friend.

In the story of the hermit's rescue of his sister from her sin the interest is very human. Only the last touch in the story speaks to us unmistakably of the deserts of Egypt from which it came. Perhaps nowhere else, after such an event, would men have sat down to discuss the destiny of the woman's soul. It was God who revealed the answer of their question to them; but we can realize how near they must have got to the love of God to have been able to receive such a revelation, when we remember that, for themselves, a whole lifetime seemed too short a space for penitence.

I

The example of St. Antony, showing how he valued a sinner who repented.

It happened that a certain brother in the community of the abbot Elias fell into sin. The brethren expelled him from the monastery and he fled to St. Antony who then dwelt on the inner mountain. The saint kept him there some time and then sent him back to the monastery from which he had been cast out. The brethren, when they saw him, immediately drove him forth again. Then, as at first, he fled to St. Antony, and said to him, "My father, they will not receive me." Then the saint was grieved, and sent to the brethren a message, saying, "A certain vessel suffered shipwreck in the sea, and all her cargo was lost. Yet with great labor the sailors brought the ship herself to land. Do you now wish to push forth into the deep and sink the ship that has been rescued? "The brethren meditated upon the message which the saint sent them. When they understood it they were greatly ashamed, and at once received again the brother. who had sinned.

II

How the abbot Besarion desired to share the reproach of the Lord, of whom they said, "He eateth with publicans and sinners."

A certain brother had sinned, and the priest ordered him to go out of the church and depart from the company of the brethren. Then the abbot Besarion arose and went out along with him, saying, "I also am a sinner."

III

How the abbot Pastor wished to deal gently with one of the Lord's little ones.

A brother came to the abbot Pastor and said, "I am working hard at the tilling of my land, for I desire to make a feast for the brethren." The abbot Pastor said to him, "Go in peace, my son, you are doing a good work." Then the brother departed joyfully, and labored yet more that he might add something to the feast he was preparing. But the abbot Anub, who had heard what was said, rebuked Pastor, saying to him, "Do you not fear God, that you have spoken thus to a brother, telling him to make a feast?" The abbot Pastor, being grieved, was silent. After two days, he sent for the brother to whom he had spoken and, Anub being present, said to him, "What was that which you asked me the other day, for my mind was wandering when I answered you?" The brother replied to him, "I told you about the tilling of my field and the harvest of it, and the feast that I was making." The abbot Pastor said to him, "I thought you were speaking of your brother who is still in the world. The making of feasts is no work for a monk." The brother was bitterly grieved when he heard this, and cried out, "I know no other good work to do, neither am I able to do any other; may I not till my farm for the sake of the brethren?" So saying, he departed. Then the abbot Anub was exceedingly sorry, and said, "My father, grant me your pardon." Pastor said to him, "Behold! I knew from the beginning that the making of feasts was no work for a monk, but according to the capacity of his mind I spoke to him. At least I excited his mind to a work of love. Now he is sad and despairing, and he will make his feast just the same."

IV

How one, through exceeding great love for his brother, suffered himself to lie, and how the fathers saw that he did well.

Two brethren once went together to a town in order to sell the things that they had made during the previous year. One of them went out to buy certain things that were necessary for them. The other, meanwhile, waited for him in the inn. At instigation of the devil this one fell into sin. When the other returned he said, "Lo, we have obtained what we wanted, let us now return to our cell. But he who had sinned replied, "I cannot return with you." The other pressed him greatly, saying, "But why can you not return." Then he confessed, saying, "Because when you were absent I fell into sin, and now it is impossible for me to go back." Then the other, being very desirous of winning and saving his brother's soul, said, and confirmed his words with an oath, "I also, while I was away from you, fell just as you did. Nevertheless let us return to our cell and repent. All things are possible with God. It is even possible that He will pardon us if we repent, and not allow us to be tormented in the eternal fires of hell." Thus these two returned to their cell. They went to the elders who dwelt near them, and casting themselves at their feet, told the story of their temptation and their sin. Whatever the elders bid them do as penance they

38

faithfully performed. The brother who had not sinned did penance for the other's sin because of the great love that he bare to him. Then the Lord looked down from heaven and beheld this mighty labour of love. After a time the whole matter was revealed by the Lord to the fathers, and they saw the great love of the brother who had not sinned, how he afflicted himself for his brother's salvation, and how the Lord had granted pardon to the sinner. "This," they said, "is that which is written. He has laid down his own life for the sake of his brother's salvation."

V

The abbot Pastor teaches a certain hermit to think of his own sins and bewail them before judging and condemning a sinning brother.

Once one of the brethren in a congregation fell into sin. Now there happened to be in that district a hermit who was renowned because for a long time he had not left his cell. To him the abbot of the congregation went and told the story of the brothees fall. The hermit, when he heard it, said, "Expel that man." So the sinning brother, driven forth from the community, went away to a desolate swamp and lamented. Now it came to pass that certain brethren on their way to the cell of the abbot Pastor heard him weeping in the swamp. They went down and found him altogether overwhelmed with grief. Filled with pity, they asked him to go with them to the cell of the abbot Pastor. He would by no means agree to go, but kept saying, "Let me stay here and die." When these brethren came to the abbot Pastor, they told him of the man whom they had found weeping in the swamp. He immediately begged them to go back again and say, "The abbot Pastor bids you come to him." When the poor man heard their words, he arose and went with them. When Pastor saw him with all the marks of his grief upon him, he arose and kissed him. Then bidding him be of good cheer, he set him down to meat. In the meanwhile he sent a brother to the hermit who had condemned the sinner, with this message: "I have heard much of you, and now for a long time have desired to see you. Now, therefore, if it be the will of God and convenient to you, I beseech you to put yourself to the toil of coming hither.' When the hermit heard these words, he said within himself, "No doubt God has revealed to him the kind of man I am, and therefore he has sent for me." Then rising up, he went to the cell of the abbot Pastor. When they had greeted each other and sat down, the abbot Pastor said, "There were two men who dwelt in one town. In the house of each of them there lay the dead body of a friend. The one of them forgot his own dead friend and the lamentation that was due to him, and leaving him unburied, went to weep at the other's funeral." The hermit when he heard these words was cut to the heart. He confessed that he had been angered at the sin of another while he forgot his own sin. Then he said, "Surely Pastor dwells in heavenly places, but I am here below on earth."

VI

How the conviction of his own sinfulness manifests itself in more gentleness towards the sins of others.

The abbot Moses said, "Unless a man is convinced in his own heart that he is a sinner,

God does not listen to his prayers." Then one of the brethren said to him, "What does it mean, this conviction in a man's heart that he is a sinner?" The old man said to him, "He who is conscious of his own sins has no eyes for the sins of his neighbour."

VII

The story of a certain brother's love for a sinner, and how he gained thereby his sister's soul.

A certain brother dwelt in a cell in Egypt who was renowned for his humility. Now he had a sister who was a harlot in the city, and was working the destruction of the souls of many men. Many times the elders exhorted him, and at last hardly persuaded him to go to her if, perhaps, he might persuade her to leave her sinful life. When he came to the town one of the citizens ran before him to the harlot's house and told her, "Behold, your brother comes to see you." She then, because she loved him, left her lovers on whom she was attending, and without even covering her head, ran to meet him. He immediately stretched forth his arms to her, and said, "My sister, my dearest sister, have pity on your own soul. Do you not know that through you many are going to perdition? How can you bear this bitter life of yours? How will you bear the torments of eternity?" She trembled exceedingly, and replied to him, "My brother, are you sure that there is salvation for me even now?" He answered her, "If you wish for it there is salvation for you." Then she fell at his feet, and besought him that he would take her with him into the desert. He said to her, "Go, then, cover your head and follow me." But she replied, "No. But let us go straightway. It is better that men should see me walking through the streets with my head uncovered than that I should go again into the place where I sinned." Then they went together, and by the way he taught her the meaning of repentance. At last, as they journeyed, they saw some men coming towards them on the road, and the brother said, "Since these men will not know that you are my sister, I beseech you go aside a little from the road until they pass." After the men had passed, he called her, saying, "Sister, let us go on upon our way." When she did not answer him, he went to look for her and found her dead, and lo! her footprints were full of blood, for she had started on their way barefooted.

When the elders heard the story they talked among themselves of whether she was saved. God in the end revealed it to one of them, that inasmuch as she had cared nothing for her body or its pain upon her journey, inasmuch as she had counted her wounds as nothing for the great longing that she had to escape perdition, that therefore, for the sake of her heart's devotion, God had received her repentance.

VIII

How an old monk was redeemed from his sin by the gentleness and patience of his disciple.

There was a certain old monk who was a drunkard. He used to weave a mat every day, sell it in a neighboring village, and spend the money he got on wine. After a while there came a younger brother, who dwelt with him as a disciple. He also wove one mat every day. The old man used to take his mat, too, and sell it, and spend the price of both on wine. Late in the evening he used to return and bring the disciple a very small piece of bread. Thus three years went by, and the young man spoke no word of complaint. At last he said within himself, "I am nearly naked, for my clothes are worn out. I am half starved for want of food. It is good that I arise and go hence." Then again he said within himself, "Whither have I to go? Better that I stay here. It was God who set me here. For God's sake, therefore, I will stay, enduring the life which I live." Immediately that he had thus resolved an angel of the Lord appeared to him and said, "You need not depart. To-morrow we shall come to you." Then the brother said to the old man, "Do not leave the cell to-morrow, I beseech you, for some friends of mine are coming to take me away." The next day, when the hour came at which the old man was wont to go down to the village, he grew impatient, and said to the disciple, "I think your friends will not come to-day. See how late it is." But the brother besought him very earnestly to stay saying that his friends most certainly would come. While he was speaking death came to him, and he slept peacefully. Then, when the old man saw that he was dead, he wept bitterly, and cried out, "Alas! alas! for me, my son! These many years I have lived carelessly; but you, in a brief time, have gained salvation for your soul by being patient." From that day forth the old man was sober, and well reported of for his good life.

IX

How the abbot Macarius by his love won for Christ the soul of a heathen priest.

Once the abbot Macarius took a journey to Mount Nitna, and, as his custom was, sent his disciple to walk some way in front of him. The young man, as he went, met one whom he recognized as the priest of a heathen temple, bearing upon his shoulders a heavy log. At once he cried out against him, saying, "Where are you going, you devil?" The priest, goaded to anger by his words, beat him and left him fainting. Then he went again upon his way. Soon he met the abbot Macarius, who said to him, "Peace be with you, toiler, peace be with you." The priest replied, "What good do you see in me that you greet me thus?" Macarius said, "I wish you peace because I see you toiling, and because you know not where you go." Then said the priest, "Your words have touched my heart. You are, indeed, a true servant of God. As for that other wretched monk who met me and insulted me, I replied to his words with blows." Then, taking hold of the feet of the saint, he said, " I shall not leave you till you teach me to be a monk." They walked together to the place where the disciple lay. Together they bore him, for he could not walk, until they brought him to the church. There the brethren were struck with astonishment to see the heathen priest in company with Saint Macarius. Nevertheless, they received him and taught him to be a monk and many of the heathen round about were converted along with him. Often afterwards Macarius used to say to them, "See how haughty words turn even good men into bad, and how true it is that loving, lowly words change bad men into good."

X

How the abbot Ammon hid a brother's sin, but warned him of his danger.

41

Once the abbot Ammon came to a certain Place to eat bread with a brother who bore an evil reputation. Now it happened that a woman had gone into this brother's cell. The inhabitants of the place were aware of it, and gathered together in great wrath to expel that brother from his cell. Hearing that Ammon was present, they asked him to go with them. As soon as the brother saw them coming, he hid the woman whom he had received in a large chest. When the crowd arrived at his cell, the abbot Ammon guessed what he had done, but for God's sake he concealed it. He entered the cell, sat down on the chest, and then bid them search. When they had looked everywhere and not found the woman, the abbot Ammon said to them, "Where now are your suspicions? God grant you pardon for them." Then he prayed with them, and bid them depart After they were all gone, he took the brother by the hand and said, "My brother, beware." So saying, he departed.

CHAPTER 6

On Humility

The Lord said: - I am lowly in heart.

- *St. Matt.* xi. 29.

It is written of Him: - He made Himself of no reputation, and took upon Him the form of a servant.

- *Phil.* ii. 7.

He came lowly, and riding upon an ass.

- *Zech.* ix. 9.

He humbled Himself, even to the death on the cross.

- *Phil.* ii. 8.

Unto the humble He revealeth His secrets, and sweetly draweth nigh and inviteth him unto Himself.

- *The Imitation of Christ,* ii. 2.

True humility, The highest virtue, mother of them all.

- *Tennyson, Holy Grail.*

PRIDE is the last and deadliest of the eight great faults which beset the feet of the hermits on their way to perfection. Over against it stands the virtue of humility, with its ultimate expression, discretion. Pride may be described as inward self-assertion. From the world's point of view, there is a right and proper kind of pride, a pride which saves men from permitting themselves even to contemplate the possibility of certain kinds of baseness. This kind of pride is what we mean by self-respect. It is the assertion of self to self. Just as the strong man asserts himself against his neighbors, refusing to be led or driven, so the self-respecting man asserts himself against himself, and, because he is determined to maintain himself for what he is, declines to be lured or goaded into ways which, for good or evil, would involve his becoming other than he is. It is perhaps impossible to draw any hard line between that self-respect which is recognized as good, and pride which is admittedly evil. Indeed, even the word "pride" itself is sometimes used, unmodified by any adjective to express a quality in man which is regarded as a virtue. This confusion of our moral judgment is the result of trying to combine the moral ideal of the teaching of Christ with the uninspired morality of even very noble men.

The hermits were perplexed with no such difficulty. To them there were no such virtues as proper pride and self-respect. All assertion of self was evil. Self-assertion against God was rebellion and sin. Self-assertion against men was the outcome of pride, its external expression. Self-assertion within was pride, in however attractive garments it might deck itself. Their judgment in the matter was absolute. They refused to recognize any kind of pride as virtuous. So it must be that many of their favorite examples of humility will strike the ordinary reader as morbid and exaggerated, and some of their heroes will not seem heroes at all, but weak creatures wanting in self-respect. For instance, the monk who groveled on the ground, beseeching pardon, while his brother beat him for a fault he had not committed, must no doubt seem to most of us to be contemptible. To the hermits who told his story he was a hero just for the same reason that makes him seem to us contemptible. He had no proper pride. He not only refused to assert himself against his brother by insisting on his innocence, but he refused to assert himself to himself; and asked pardon for what he had not done. To the devil also, who had plotted the separation of the brethren, this man seemed to be a hero, one so near to God as to be unconquerable. It does not really matter whether the story is literally true or not. Our consciences recognise that what the story relates would always really happen. The devil could not but fly defeated from such humility as this man showed.

Most of the stories and sayings, however, make no such strange demands upon our moral sense. We recognize gladly the lofty teaching of the story of the hermit to whom the desired revelation came only after he had humbled himself. We readily admire, even if we are slow to imitate, the humility of Arsenius, who was not ashamed to accept spiritual teaching from an ignorant peasant. Nor when we remember the dangers which beset the hermit's path shall we be astonished at the vision of St. Antony, and the voice which came to him praising humility. There were dangers of which most men know nothing, like that of the monk to whom the devil came disguised as the angel Gabriel, or that which beset St. Ammon when men asked him to judge between them.

The thought of humility and the desire of it was very constantly present to the hermit's mind. I do not find on any other subject so many brief, and as one may say proverbial, words as on humility. In some of these the thought is so condensed as almost to defy intelligible translation, but I am sure that a careful study will reveal in each of them some spiritual thought which will well repay the labor of pursuing it.

I

Of the great safety of being humble.

St. Antony tells how once in a vision he beheld all the snares of the evil one spread over the whole earth. When he looked upon them and considered their innumerable multitude, he sighed, and said within himself, "Who is able to pass safely through such a world as this?" Then he heard a voice, which answered him, "The humble man alone can pass safely through, O Antony. In no way can the proud do so."

II

A story of how a certain one escaped one of the snares of the devil through humility.

The devil once appeared to a certain brother transformed into the likeness of an angel of light. He said, "I am the angel Gabriel, and I am sent unto thee." The brother, though he doubted not at first but that he saw an angel, yet out of his humility made answer, "Surely you are sent to some other one and not to me, for I am altogether unworthy to have an angel visitor." Then the devil, being astonished and baffled, departed from him.

III

The humility of the abbot Arsenius who once dwelt in the emperors court.

The abbot Arsenius was one day talking with an ignorant peasant monk about spiritual thought. Another monk saw him doing so, and said to him, "How is it, Arsenius, that you, who know both Latin and Greek, consult this peasant about his thoughts?" Arsenius answered him, "I do, indeed, know Latin and Greek, which contain the wisdom of this world, but I have not yet succeeded in acquiring even the alphabet of what this peasant knows. His wisdom is of another world."

IV

How a brother once obtained a spiritual benefit as a reward for his humility. It is related of a certain brother that he once persevered in fasting for seventy weeks. This he did desiring to obtain a divine illumination on the meaning of a certain passage in Holy Scripture. Nevertheless, though he so fasted and desired, God hid the matter from him. Then, at last, he said within himself, "See, I have undergone great toil and am nothing profited. I shall go to one of the brethren, and inquire of him what this word of Scripture may mean." So saying, he went out and closed the door of his cell after him. Immediately then an angel met him and said, "The seventy weeks of your fasting have not brought you near to God that you should know His mind. Now, however you have humbled yourself in going to inquire of your brother. Therefore I am sent to reveal to you what you desire to know." Then the angel opened to him the matter about which he was perplexed, and departed from him.

V

How a divine and eternal reward awaits those those humility has taught them to regard their own labour as nothing.

A certain father said, "He who labors and considers that by his labor he has accomplished or effected anything, has already, even here, received the reward of all that he has done."

45

VI

The way in which a certain brother learnt and practiced humility.

There was a certain brother who belonged to a high family, as this world reckons rank and grandeur. He was the son of a count, and was extremely wealthy; also he had been well educated as a boy. This man fled from his parents and his home, and entered a monastery. In order to prove the humility of his disposition and the ardor of his faith, his superior ordered him to load himself with ten baskets and to carry them for sale through the streets of the city. If anyone should want to buy them all together he was not to permit it, but was to sell them each to a separate purchaser. This condition was attached to his task in order to keep him the longer at work. He performed his task with the utmost zeal. He trampled under foot all shame and confusion for the love of Christ and for His name's sake. He was not perturbed at all by the novelty of his mean and unaccustomed work. He thought neither of his present indignity nor of the splendor of his birth; he aimed only at gaining through obedience the humility of Christ, which is the true nobility.

VII

Words of the hermits concerning humility.

Evagrius said: "The beginning of salvation is to despise yourself."

Pastor said: "A man ought to breathe humility as his nostrils breathe the air."

Another said: "Humility is that holy place in which God bids us make the sacrifice of ourselves."

Syncletica said: "As no ships can be built without nails, so no man can be saved without humility."

Hyperichius said: "The tree of life is on high. Man climbs to it by the ladder of humility."

Another said: "It is better for a man to be conquered by others on account of his humility, than to be victorious over them by means of pride."

Another said: "May it ever be my part to be taught, and another's to teach."

Cassian said: "It is never said of those who are entangled in other sins that they have God resisting them, but only 'God resisteth the proud.'"

Motois said: "Humility neither is angry nor suffers others to be angry."

The abbot John the Short said: "The door of God is humility. Our fathers, through the many insults which they suffered, entered the city of God."

He also said: "Humility and the fear of God are pre-eminent over all virtues."

VIII

How one yearned for perfection, and God taught him to be humble.

There was a certain old man who dwelt in the desert, and it seemed to him that he had learnt the perfection of all the virtues which he practiced. So he prayed to God, saying, "Show me what is yet lacking for the perfection of my soul and I will accomplish it." Then God, who wished to teach him humility of mind, said to him, "Go to the leader of a certain congregation of monks, and what he bids you, that do." At the same time God spoke to that leader of monks and said, "Behold, the solitary of whom you have heard comes to you. Bid him take a whip and go forth to herd your swine." The hermit arrived, knocked at the door, and entered. When they had saluted each other and had sat down, the hermit said, "Tell me, what shall I do to be saved." The other, doubting within himself, replied, "Will you do what I bid you?" The hermit said, "Surely, yes." Then said the other, "Lo! Take this whip and go forth and herd my swine." While the hermit drove the swine out to their pasture there came by some men who knew him, and they said, "Do you see that famous hermit of whom we heard so much? He must have gone mad, or some demon possesses him. Look at him feeding swine." All this the hermit endured patiently. Then God saw that he had learnt humility, and was able to bear the insults of Therefore He bid him return to his own place.

IX

How a certain elder shrank from being praised, and rejoiced when he was despised.

A certain old man dwelt in the lower part of the desert, at peace, in a cave. A religious man from a neighboring village used to bring him what he wanted. It happened that this man's son fell sick. With many prayers he besought the old man to come to his house and pray for the child. At length he prevailed with him, and running home, cried out, "Prepare for the coming of the hermit." When the people of the village knew that he was coming they went out with torches to welcome him as if he had been some prince or governor. The hermit, as soon as he perceived how they meant to greet him, stood upon the river-bank, and taking off his clothes, went naked into the water. When the man who was accustomed to minister to him saw this he was greatly ashamed, and said to the villagers, "Return to your homes, for our hermit has lost his senses." Then going to the old man, he said, "My father, why have you done this? All those who saw you are saying, 'That old man is nothing better than a fool.'" The hermit replied to him, "That is the very thing I wished to hear."

X

How St. Ammon became a fool for Christ's sake.

This story is told of the abbot Ammon. Certain men came to him asking him to judge in a contention which they had. He, however, would not, and put them off. Then a woman said to another woman who stood near her, "The old man is silly." Ammon heard her words, and calling her to him said, "For very many years I have toiled in various solitary places to attain that silliness at which you scoff Is it likely now that I shall be content to lose it because you taunt me.

XI

The abbot Pastor's description of humility.

The abbot Pastor was once asked by a monk: "How ought I to conduct myself in the place where I dwell?" He answered, "Be cautious as a stranger among strangers. Wherever you are, never seek to have your own opinion prevail or your word influential. So you may have peace.

XII

How the devil was vanquished by the great humility of one of the brethren.

There were two brethren, relatives according to the flesh, and bound to each other yet more closely by the spiritual purpose of their devotion. Against them the devil laid a plot that he might separate them the one from the other. Once, towards evening, the younger of the two, as he was wont, lit their lamp and put it on its stand. Through the malice of the devil the stand was overturned, and the lamp went out. By this means the devil hoped wickedly to entrap them into a quarrel. The elder of the two, growing suddenly angry, struck the younger fiercely. But the younger fell humbly on the ground and besought, saying, "Sir, be gentle with me, and I will light the lamp again." Then, because he gave back no angry word, the evil spirit was filled with confusion, and departed from their cell. That same night he told the chief of the devils the story of his failure, saying, "Because of the humility of that brother who fell upon the ground and begged the other's pardon I was unable to prevail against them. God beheld his humility, and poured His grace upon him. Now, lo! it is I who am tormented, for I have failed to separate these two or make them enemies."

XIII

Another story of a devil vanquished by humility.

There was a certain hermit renowned among the monks. It happened that there once met him a man possessed by an evil spirit, who struck him violently upon the cheek. The old man straightway turned to him the other cheek, that he might smite him upon it also. The devil was not able to endure the flame of his humility, but immediately departed from him who was possessed.

48

CHAPTER 7

On Discretion

The light of the body is the eye: if therefore thine eye be single, thy whole body shall be full of light. But if thine eye be evil, thy whole body shall be full of darkness.

- *St. Matt.* vi. 22, 23.

Some persons, inexperienced in the grace of the devout life, have overthrown themselves, because they attempted more than they were able to perform, not weighing the measure of their own weakness, but rather following the desire of their heart than the judgment of their reason.

Better it is to have a small portion of good sense with humility and a slender understanding, than great treasures of knowledge with vain self-complacency.

- *The Imitation of Christ,* iii. 7.

IT is discretion which enables a man to judge rightly in matters concerning his religious life. We must remember that during the earlier stages of the monastic movement the monks were not very eager about doctrinal orthodoxy. It is true that even to be suspected of heresy seemed horrible and quite unbearable, yet their main interest was centered in the problem of how to live the Christian life rather than in the definitions of the Christian faith. It is therefore in the decision of practical questions that they conceived discretion to be useful. They made no claim to special illumination in matters of the faith.

The teaching of history confirms the belief, which most of us arrive at by experience, that the character of religious enthusiasts is likely to be marred, not only by narrowness and bitterness, but by exaggerated rigorism and by spiritual pride. The hermits saw a danger of ultimate shipwreck for the religious soul in which these faults were suffered to find a home. Safety lay in cultivating the virtue of discretion. There are, of course, other safeguards. The genial laughter with which humanity's broad common sense greets exaggeration and affectation saves the world from being overrun with militant faddists of various kind. For most men the laughter of their friends is the best cure for religious and moral foolishness. It is unfortunate that the genuine enthusiast is to some extent removed from the fear of being laughed at. The hermits, especially, stood clear from the influence of the world's opinion. An essential element in their position was that they had learned to say with St. Paul, "To me it is a very small matter to be judged of you or of man's judgment." To many a hermit it came to be actually a source of satisfaction to be laughed at as a fool. The greatest among them did not shrink from using ridicule as a weapon in combating exaggeration and pride. Perhaps a sense of humor is the last thing we should expect to find in a hermit. Yet we find not only that they could appreciate the laughable side of the faults they combated, but that they were quite ready to enforce the teaching of common sense by extremely practical forms of ridicule. It will not be denied that they were wise. The Latin poet, who claimed the right of speaking the truth while he laughed, was too modest. He might have asserted that certain kinds of truth can hardly be spoken at all except while laughing. The exaggerations and eccentricities of faddists tend, however, to grow aggravated until we learn to call them fanaticism and insanity. Then they can no longer be met or cured by laughter. It is against the danger of moral shipwreck from these faults that the hermits sought safety in learning discretion. One of the most striking features in the characters of the greatest among them, is their entire sanity. In spite of the strangeness of St. Antony's way of life and the severity of St. Macarius' asceticism, no one will be found who wants to laugh at them. Stories of lesser enthusiasts move men to smile even while they love. The great Egyptian hermits may be hated, indeed, but they cannot be regarded as either ridiculous or fanatical. The saving grace was discretion.

Discretion, as the hermits conceived it, differed from common sense or an appreciation of the ridiculous, in that it was not a natural faculty but a virtue. A faculty is a gift whose possessor is to be envied. A virtue is an attainment. He who has it, has attained it. He who lacks it, is a failure. The hermits regarded discretion as the final glory of a perfect character. Sometimes it saved them from excessive asceticism. After St. Antony preached his first great sermon to the monks on his outer mountain, moderation was recognized as a fruit of discretion. There are stories which inculcate fearful warnings against extremes of fasting and labor. It brought with it a keen spiritual insight which saved the monk from hating or despising the world which God made very good, or claiming for himself an imaginary freedom from the divinely imposed necessity for labor. It saved him, too, from diabolic deceptions. It is by a form of discretion, amounting almost to inspiration, that the hermit recognizes the devil who comes to him disguised as an angel of light. This is its highest achievement. Discretion is the eye of the soul. It is of it, according to St. Antony, that the Lord taught when He said, "The light of the body is the eye. If, therefore, thine eye be single, thy whole body shall be full of light." There is only one way of attaining this perfect vision, and that is through humility. Indeed, discretion and humility are the same virtue. Humility is the root and stem, discretion is the flower and fruit. It is quite obvious that this is so. Only the man who has too good an opinion of himself will break out into the absurdity of eccentricity and singular exaggerations. Only the man whose soul is puffed up with spiritual pride is in any real danger of mistaking the devil for an angel of light, charged with a special revelation to him. It is by steadily maintaining a low opinion of himself that a man may hope to achieve that singleness of eye of which the Lord speaks, the ability to distinguish with swift and certain glance between what is really good and those subtler forms of evil which dress themselves in virtue's clothing.

I

A discourse of St. Antony, wherein is explained the meaning and the value of discretion.

Often men are most strict in fasting and in vigils. Often they nobly withdraw into solitude and aim at depriving themselves of all their goods so that they do not suffer even one day's supply of food or a single penny to remain to them. Often they fulfill all the duties of kindness with the utmost devotion. Yet even such men are sometimes suddenly deceived. They cannot bring the work they have entered upon to its fitting close, but bring their exalted fervor and noble manner of life to a terrible end. In these men, though the virtues I have mentioned abound in them, yet discretion is wanting, and they are not able to continue unto the end. There is no other reason for their falling away than that they have not obtained discretion, that spiritual wisdom which, passing by excess on either side, teaches a monk to walk always along the royal road. It does not suffer him to be puffed up on the right hand of virtue, that is, from excess of zeal, in foolish presumption, to transgress the bounds of due moderation. Nor does it allow him to become slack and turn away to vices on the left hand, that is, under pretext of duly managing the body, to become lukewarm. For it is discretion which is termed in the gospel the "eye" and "the light of the body" according to the Savior's saying, because as it discerns all the thoughts and actions of men it sees and overlooks all things which should be done. But if in any man this be "evil," that is, not fortified by sound judgment and knowledge, or is deceived by some error or presumption, it will make the whole body "full of darkness." It will obscure all our mental vision, and our actions will be involved in the darkness of vice and the gloom of unpeacefulness. No one can doubt that when the judgment of our heart goes wrong and is overwhelmed by ignorance, our thoughts and deeds must be involved in the darkness of still greater sins.

II

A story of the abbot John the Short: how he fell into the sin of presumption through lack of discretion, and afterwards was saved.

They tell this story about the abbot John the Short. Once he said to one of the brethren who was his senior, "I wish to be as the angels are, free from all care, doing no work, but ceaselessly praising and praying to God." Then casting off his raiment, he departed into the wilderness. After a week had passed, he returned to his brother and knocked at the door of his cell. Before he opened to him, the brother asked, "Who art thou?' John replied, "I am John." Then the brother answered him and said, "Not so, for John has become an angel, and no longer has intercourse with men." He, however, continued knocking, and crying out, "Indeed, I am he." The other, however, would not open the door, but left him suffering there. At last he opened the door and admitted John, saying to him, "If you are a man, need is for you to work that you may live. If you are an angel, why do you seek entrance to my cell?" John then, being truly penitent, replied, "Pardon me, O brother, for I have grievously sinned."

III

The abbot Evagrius commeds discretion in advising that all things be done moderately and at ftting seasons.

The abbot Evagrius said: Reading and watching and prayer are good for the slothful spirit and the wandering mind. Fasting and toil and carefulness will tame lust though it burn in us. The singing of psalms, together with patience and tenderness, will conquer wrath and bring peace in troubled times. Yet must all these be practised at due times, and all within the bounds of moderation. For he who exercises himself in these ways inopportunely and excessively may indeed profit for a little while, but after a short time will be harmed, not helped, by them.

IV

How the abbot Lucius rebuked certain brethren who showed that they lacked discretion, and taught them a better way.

Certain brethren once came to the abbot Lucius, and the old man asked them, "What work are you wont to do?" They said, "We do no work, but, according to the saying of the apostle, we pray without ceasing." Then said the old man, "Do you never eat?" And they replied, "Truly, we do eat." Then Lucius said, "And who does your praying for you while you eat?" They were silent. Then he asked them "Do you never sleep?" When they confessed that they slept, he asked, "And who does your praying for you while you sleep?" They could find no answer to give to him. Then he said, "I see that you do not perform what you boast. I will show you how to pray without ceasing. Sit working in the morning up to the accustomed hour; weave mats and make baskets. Meanwhile keep praying in these words: 'Lord, according to thy mercy pardon my offences and do away with my iniquity.' When you have finished a few baskets sell them for money. Give a portion to the poor, and keep the rest to buy your food. When, then, you eat or sleep, the poor whom you relieve are filling in the gaps in your ceaseless round of prayer."

V

The abbot Pastor teaches discretion to a brother who repented truly of his sins.

A brother asked the abbot Pastor, "I have committed a great sin. Shall I do penance for three years?" Pastor replied to him, "That is too long." Then the brother said, "Do you advise one year?" Again Pastor replied, "That is too long." Those who were standing by asked, "Are forty days sufficient?" Pastor said again, It is too long." Then he added, "If a man repent with all his heart, and fully determine not to commit again the sin which he deplores, God will receive his repentance though it endure but three days."

VI

Of a wandering brother who lacked discretion, being puffed up with spiritual pride.

A certain wandering brother came to the monastery of the abbot Silvanus. He saw the brethren working, and rebuked them, saying, "Why do ye labor for the meat which perisheth? Mary chose the good part." Then said the abbot Silvanus to his disciple Zacharias, "Give this brother a book to read and put him into an empty cell." At the ninth hour the brother looked out and gazed along the path to see if any man was coming to call him to a meal. After a while he went to Silvanus, and said, "Do not the brethren eat to-day?" The abbot confessed that they had already eaten. Then said the brother, "Why did you not send to call me?" Silvanus answered him, "You are a spiritual man. You have surely no need of such food as we eat. We, indeed, are but carnal; we must eat. We labor, but you have chosen the good part. You read all day, and have no wish to receive carnal food."

VII

Of discretion in prayer. Certain brethren asked St. Macarius how they ought to pray. He answered them, "There is no need of much speaking in our prayers. Stretch out your hands and say, 'Lord, have mercy upon me as Thou wilt and as Thou seest best.' If your mind is disquieted, then say, 'Help Thou me.' He knows well what is best for us. Of His own will He grants us mercy."

VIII

How discretion taught Nathyra to alter his rule of life according to the circumstances amid which he found himself.

The abbot Nathyra, the disciple of Silvanus, when he lived as a hermit in his cell, adopted a very moderate rule of life, allowing himself all that was necessary for the welfare of his body. Afterwards, when he became a bishop, he used a much severer discipline. One of his disciples asked him, saying, "Master, when we dwelt together in the desert you used not thus to crucify yourself; why do you do so now ?" The bishop said to him, "My son, there in the desert we had solitude and quietness and poverty; therefore I so regulated my bodily life that I should not grow weak, but be able to strive for those graces which I desired. Here in the world are many temptations to excess of every kind; moreover, here there are many to warn me should I overtax my strength with fasting. I live austerely here, lest I should let slip the hope of perfection which led me to become a monk."

IX

The abbot Agathon gave evidence of his discretion by avoiding all extravagance.

The abbot Agathon so managed his life and his affairs that discretion appeared to govern everything he was or did. This was the case not only in great matters, such as the labor which he performed, but even in the details of his dress. Thus he wore such clothes as never could strike anyone as either particularly good or particularly poor.

X

How one was preserved from a snare by discretion. They tell about a certain old man that sometimes in his struggles against temptations he saw the devils, who surrounded him, with his bodily eyes. Nevertheless, he despised them and their temptations. Seeing that he was being vanquished, the devil came and showed himself to the old man, saying, "I am Christ." But when the old man beheld him, he shut his eyes. Then the devil said again, "I am Christ; why have you shut your eyes?" The old man answered him, "I neither expect nor wish to behold Christ in this present life. I look to see Him only in the life beyond." Hearing these words, the devil straightway vanished from his sight.

<h2 style="text-align:center">XI</h2>

The story of another who was saved by discretion from an illusion.

There was another old man whom the demons wished to seduce. They said to him, "Do you wish to behold Christ?" He replied to them, "May you be accursed for the words you speak. I believe my Christ when He says to me, 'If anyone shall say unto you, Lo, here is Christ or lo there, believe him not.'" When they heard him answer them thus the devils immediately vanished.

<h2 style="text-align:center">XII</h2>

A way in which a man may order his life wisely.

A certain brother asked the abbot Antony, "What shall I do that I may please God?" The old man replied, "Keep these commandments which I give you. Wherever you go, have God always before your eyes. Whatever work you do, set before yourself an example from the Holy Scriptures. Wherever you dwell, be not hasty in removing thence. Stay patiently in the same place. If you guard these three precepts without doubt you will be saved."

CHAPTER 8

On the Necessity for Striving

The kingdom of heaven suffereth violence, and men of violence take it by force.

- *St. Matt.* xi. 12 (R.V.).

Be thou therefore ready for the conflict, if thou wilt have the victory. Without a combat thou canst not attain unto the crown of patience. Without labor there is no arriving at rest; nor without fighting can the victory be attained.

- *The Imitation of Christ*, iii. 19.

NO prize worth the having can be obtained without effort. The worthier the prize, the greater and more continuous must be the effort. The noblest of all prizes which a man can seek is that to whose attainment the hermits dedicated their lives, the perfection which is in Christ Jesus. We are therefore in no way surprised when we find that they were forever bracing themselves for strenuous effort. Their experience in this respect is common to all Christians, for everyone who hungers and thirsts after righteousness must so brace himself. No one who is in any real sense a follower of Jesus Christ can fail to be aware that the world is always luring him from the narrow way, and that there is something within him which responds eagerly to the enticements spread beside his path. Hence comes the necessity for watchfulness and effort. But the Christian is not only like a traveler who goes along a toilsome road beset with inducements to leave it; he is like a soldier on the march through an enemy's country; he is surrounded by beings hostile to his progress; his journey involves not only effort, but strife, and that against powers, personal, crafty, and desperately malevolent.

The intensity of the hermits' realization of this strife is one of the most striking features of their conception of the religious life. Perhaps no one since the days of the apostles ever realized as the hermits did the meaning of the Lord's saying about "the strong man armed who keepeth his house." The neophyte entered upon his ascetic life with the full consciousness that he would be assailed in every way which diabolic ingenuity could devise. I have not translated any of the stories which tell of the devil's attempts to terrify the hermits with frightful sounds, or of the physical violence which he did to them. Such things lie too far from the experiences which seem possible to us to be of much profit. We must however remember that the hermits expected and endured such assaults if we are to appreciate their conception of the Christian strife. Other stories like those which tell of the suggestion of evil thoughts and the making of opportunities for strife among the brethren we can very easily understand. When we do so, when we appreciate the ceaselessness of the strife and the weariness involved in it, we shall be in a position to admire the enthusiasm which recognized the value of the struggle. For the hermits regarded the strife itself as an indispensable discipline apart altogether from the value of the particular virtue they fought for. It is inspiring to think of the man who desired that evil thoughts should continue to assault him, rather than that he should be freed from them, because he felt that the constant strife was better for his soul than quietness. In the same spirit is conceived the exhortation of the elder to John the Short. It was no doubt something to enjoy the peace which followed his escape from envy and evil thoughts. Yet the advance of the soul towards God was felt to have stopped when the strife ceased. The thought of the hermits in the matter is perhaps best expressed in Pastor's strange interpretation of the Lord's words, "He who hath no sword let him sell his garment and buy one."

This great conception of the value of strife and struggle in the formation of the perfect man serves to explain much that is otherwise puzzling and perhaps distasteful to us in the asceticism of the hermits. The monk who moved his cell yet further from the well out of which he fetched his water will not seem to be a fool when we understand that the satisfaction of his bodily thirst was a small matter compared to the opportunity of self-conquest which his journeys through the heat afforded him. The angel who counted his footsteps was in reality reckoning up the progress of his soul towards perfection, and not merely the miles his body traveled. From another point of view, we are able to at least sympathize with extremes of ascetic practice when we read the Abbot Achilles' parable about the trees and the axe. The complete conquest of the body seemed to the hermits a primary necessity in living the religious life, because the body furnished the weapons which the devil used with most fatal effect.

From these two convictions of the value of strife in itself, and of the constant readiness of the adversary to seize on any weak point in the soul, there naturally came the stern doctrine that peace on this side of the grave is only to be found by the utterly apostate. In perfection, indeed, there is perfect peace; but perfection, as the hermits knew well, is not attainable on earth. In complete abandonment of the soul to evil there is also peace. Between these two, that is, throughout the whole region of Christian life, there can be nothing but strife, and strife which grows harder and not milder as every successive victory is won.

I

How the abbot John learnt the lesson that inward strife is better than inward peace.

The abbot Pastor relates of John the Short that he once prayed, asking God to take away from him all passion. God granted his prayer; and he, being free from envy, anger, and all evil thoughts, was at peace. In his great gladness he went to a certain elder, and said to him, ,Behold in me a man who has no strife nor contests. I am altogether at peace." But the old man, being grieved for John's sake, replied to him, "My son, go, ask the Lord to grant you occasion for strife. There is no way in which the soul advances towards God but by striving." Then John, knowing in himself that this was true, did as the old man bade him. Afterwards, when the necessity for constant strife came back upon him, he never again prayed that it should be taken away from him. Always be made this petition "Lord, give me grace to conquer in the strife."

II

A story setting forth how toil in itself is for the soul of him who desires to enjoy the kingdom of God.

There was a certain old man dwelling in the desert whose cell was above two miles distant from any water. Often when he went to draw water, and the sun shone hot on him, he grew weary. Once, as he went, he said to himself, "There is no need for me to endure all this labor. I shall go and dwell nearer to the water." As he so spake he turned and saw one following him who seemed to mark his footsteps. The old man asked him, "Who are you?" The stranger answered, "I am an angel, and the Lord sent me to count your footsteps and give you your reward." When the old man heard this he remembered that he had not come out into the desert for the sake of ease, but to travel on the narrow way that leadeth unto life. Then he became yet bolder in heart and more violent, and set his cell even further from the water.

III

The abbot Pastor's strange interpretation of a saying of the Lord.

The abbot Pastor said, "It is written in the gospel, He who has a coat, let him sell it and buy a sword. This word is to be understood by us in this manner: He who has peace let him cast it away, and in its place take unto himself strife. Now our strife is against the devil."

IV

A saying of the abbot Serenus showing that the strife is severest for those who are furthest advanced towards the kingdom of heaven.

We know well by our own experience and the testimony of the Fathers that devils have not the same power against us which they had formerly in the days of the first anchorites, when there were only a few monks living in the desert. This is because of our carelessness which makes them relax somewhat of the violence of their first onslaught. They scorn to attack us with the same energy with which they formerly raged against those most admirable servants of Christ.

V

A parable of the abbot Achilles, showing how our strife is not only against the powers of evil which are without, but also, even chiefly, against the evil that is within.

A certain brother said to the abbot Achilles, "How is it that the demons have power against us?" The old man answered him thus: "The trees of Lebanon said, 'How great we are and high! Yet we are cut down with a very small axe. Yes, and of the axe which cuts us down the greater part is wood, and comes from us. Let us therefore give no part of ourselves, and the axe will have no power against us.' Soon there came some men seeking timber, and they made a handle for their axe out of these very trees in spite of their boasting. So the trees were cut down. Now the trees are the souls of men. The handle of the axe is man's evil will. So we are cut down by means of the evil that is within us."

VI

Of one who, as a good soldier of Jesus Christ, did not shrink from the conflict.

The disciple of a certain holy old man was once attacked by a spirit which tempted him. By the grace of God he fought valiantly against the vile and impure thoughts of his heart. He used the discipline of fasting. He prayed often. He worked diligently and vehemently with his hands. The holy old man beholding his labor and strife, said to him, "If you wish it, my son, I will pray to the Lord and ask Him to remove this adversary away from you." The disciple, however, replied to him, saying, "I perceive, my father, that although I am enduring what is hard, yet good fruit is being perfected in me. By reason of the temptation which besets me I fast more than if I were at peace. I am more steadfast in waiting. I am, as I think, more earnest in prayer. I beseech you, nevertheless, that you pray for me and seek the mercy of God for me. Ask that I may be given valor to endure and to fight according to God's will." Then the old man was filled with joy, and said, "Lo! now I know, my son that you understand this spiritual conflict, how it works in you for the perfecting of your eternal salvation."

VII

Why no man may dare to think within himself 'I have conquered, and need strive no more.'

A certain old man came to another and said, "I, indeed, am already dead unto the world." But the other, seeing the danger in which he was, thus warned him, "Be not ever sure of yourself while you remain in the body. Although perhaps you may say, 'I am dead unto the world,' yet there is one who is by no means dead to you even your adversary the devil. Surely innumerable are his evil ways, and immeasurable is his craftiness."

VIII

Of toil and peace.

Isidore, a priest in Scete, said once to the brethren who were gathered round him, "Brethren, was it not in search of toil and hardship that we came hither? Behold, I find here no sufficient toil. I shall therefore gird myself, and go elsewhere and find toil. Then I shall also find peace."

IX

How toil and patience are the means of spiritual gain.

A certain elder said, "We often fail to advance because we know not the conditions of our strife, nor have we patience to complete the work we have begun. No virtue can be attained without toil."

X

How no man must cease from striving until he has attained perfection or ceased to wish for it.

A certain brother used often to go to the abbot Sisois and ask advice from him, saying, "My father, what shall I do, for I have fallen into sin?" Sisois replied, "Rise out of your sin." Again the brother came with his confession, saying, "I have fallen into sin again." The old man said to him, "Then again you must rise from your sin." Very often the brother came to him, saying, "I rose again, indeed, but again and again I have fallen." Still Sisois gave him the same advice, "You must not cease to rise from your sin again and again." At last the brother said to him, "My father, how long shall I go on rising again from my sin? Tell me this." The old man said to him, "Until you are at rest in the perfect performance of what is good, or have found quietness in complete bondage of evil."

XI

We must not think that even repeated victory over any fault frees us from the necessity for strife against it.

There was a certain old man who dwelt for fifty years in the desert. He neither tasted bread, nor even drank enough water to satisfy his thirst. At last he said, "I think I have conquered utterly - yea, slain - the sins of avarice and vainglory." When the abbot Abraham heard that he had spoken these words, he came to him and asked if it was true that he had so spoken. He confessed that it was true. Then Abraham said to him, "Suppose, now, that you were walking along the road and you saw a pile of stones and broken bricks, and suppose that you saw in the midst of them a lump of gold, are you able to look upon it just as you look upon the stones and bricks?" The old hermit answered, "No. I should feel that it was precious, but I should fight against the thought." Then said the abbot Abraham, "See, therefore. Avarice still lives in you, but you have fettered it." Again the abbot Abraham spoke to him, "Here is a man who loves you well and praises you. Here is another who hates you, and is for ever slandering you. If both of them come to you, can you look upon both of them with the same affection?" The old hermit answered him, "No. I cannot do this at once, but I should struggle with myself until I felt that I loved him whom at first I did not love." Then Abraham said, "See, now; your passions are yet alive in you, but they are bound with holy bands."

XII

How we must ever be ready to do violence to ourselves.

A certain elder was once asked, "What is the meaning of this which is written: 'Strait is the gate and narrow is the way which leadeth unto life'?" He answered, "The strait and narrow way is this: that a man do violence to his thoughts and destroy his own will for God's sake. This is what we are told the apostles of whom it is written: 'Lo, we have left all and followed Thee.'"

XIII

How in this life it is only possible to escape from strife by yielding entirely to all temptation.

A certain brother said to one of the elders, "In my life there is no strife. My soul is at peace." The elder said to him, "If that be so, you are like a wide-opened door. Whatever likes can enter into you, whatever likes can go out. You know not what is happening in your heart. For if you hold your heart's door fast, and keep it shut so that on refuse entrance to all evil thoughts, then you will see them standing without and feel that they are fighting against you.

XIV

How the life of a monk is a life of ceaseless strife.

The abbot Macarius once said to the abbot Zacharias, "Teach me wherein a monk's life consists." Zacharias replied, "Do you, my father, ask this question of me?" "I am fully determined to ask you," said Macarius, "for there is One who is spurring me on to do so." Then Zacharias said to him, "In my opinion, my father, he is truly a monk who in all things does violence to himself."

CHAPTER 9

On Fasting

And when He had fasted forty days and forty nights, He was afterward hungry.

- *St. Matt.* iv. 2.

It is possible to be saved without virginity. It is not possible to be saved without humility. Without humility (I dare even to say this) even the virginity of Mary would not have pleased God.

- *St. Bernard, 1st Homily in praise of the Virgin Mother.*

Sackcloth is a girdle good, Oh, bind it round thee still. Fasting, it is angels' food, And Jesus loved the night air chill; Yet think not prayer and fast were given To make one step 'twixt earth and heaven.

- *Lyra Apostolica*, xxxvi.

THE strife which the monks felt to be a necessary condition of all spiritual advance took place in two regions. There was strife against the body - the struggle with physical needs, desires, and passions. There was also the struggle against infirmities and failings of the soul - spiritual strife. In each region the strife is, strictly speaking, an asceticism, that is to say, an exercise undertaken with the object of attaining some further end. In the case of the physical asceticism of the hermits it is especially necessary to understand the meaning of the words we use and the real nature of the practices described. Asceticism (asksis) means an exercise, and an exercise is an entirely useless and meaningless thing unless it is undertaken with a view to something to be gained by its use. When St. Paul speaks of "exercising" himself he says that he does so in order to have a conscience void of reproach. In exactly the same way the monks practiced exercise, asceticism (asksis), not as if the things they did were in themselves good, but simply as a means to the attainment of that perfection which they desired.

The most striking form which the physical asceticism of the hermits assumed was fasting. There were other forms, but fasting was the most esteemed, and it is of fasting that we read most in the stories of their lives. There are in the annals of Egyptian monasticism some instances of terribly severe and prolonged fasts. There were hermits who ate only once every two or three days. A common practice was to eat nothing until after sunset. There was no attempt, at all events in Lower Egypt, to establish anything like a uniform rule on the subject of fasting. It was recognized that the capacity for fasting varied greatly in different individuals. One man might eat what seemed to be a great deal, and yet truly fast. Another might eat very little, and yet be a glutton. So far as the advice of the greatest Fathers can be said to form a rule, it may be expressed in the words - "Do not eat to satiety." In the spirit of this advice each hermit regulated the time of his own meals and the quantity and quality of his food as seemed best to himself.

The end which the hermits hoped to attain by fasting was the subjugation of the lusts of the flesh. The hermit who disdained the exercise of fasting was compared to a horse without a bridle. How far the hermits were from regarding fasting as an end in itself, or even as invariably the best means for overcoming fleshly lusts, may be seen from the fact that young men were sometimes advised to eat more and fast less, so as to obtain more strength to resist the attacks of their spiritual enemies. Apart, however, from the practice of fasting as an asceticism, an exercise undertaken for a purpose, the hermits fasted in simple obedience to the Lord's teaching and in sympathy with His fasting. This is part of their whole conception of the religious life as a literal imitation of Christ.

Fasting, being a merely physical exercise, is regarded always by the hermits as a practice which ought to be discontinued directly it interfered in the smallest degree with the attainment of a virtue or the fulfillment of a higher kind of duty. Thus, if success in fasting led a man into danger of becoming proud or vainglorious, it was better for him to eat, even to eat flesh. A hermit, whose severe fasting led him to envy a brother whose conditions of life were pleasanter, had better eat flesh and drink wine than fall into such a sinful state. In the same way it was felt to be better for a man to break his rule of fasting than to assert himself by keeping it when others in his company wished to eat. Active charity, such as manifests itself in hospitality to strangers was always to be preferred before fasting. It might happen that a hermit, whose ordinary observance was very strict, would break his fast even seven times in one day if seven separate strangers came to his cell demanding entertainment. In so doing he was right, for the lower duty, of fasting according to his rule, had only given place to a higher one, love showing itself in hospitality.

Sometimes it seems as if, through the exercise of fasting, the hermits actually attained to such a conquest of the flesh that its needs and demands no longer interfered with spiritual communion with God. Thus we read of solitaries who forgot to eat amid the rapture of a bliss only to be compared to the bliss of angels. We read, too, of men whose talk on spiritual matters became so absorbingly interesting that the needs of their bodies disappeared from their consciousness, even though their meal was spread ready before them.

I

How the spirit of love may loose the obligation of a fast, and yet where love makes no call on us the days of fasting ought to be observed.

The abbot Silvanus came one day with his disciple Zacharias to a certain monastery. The brethren who dwelt there besought them to eat something before they departed. They willingly received the food placed before them, lest they should grieve the brethren who offered it. Afterwards they departed. As they journeyed they came to a pool of water, and Zacharias wished to drink of it. Silvanus rebuked him, saying, "This is a fast day. You ought not to drink." He replied, "But, my father, have we not already eaten and broken our fast?" "My son," said Silvanus, "that eating was for the sake of the brethren, because we loved them. Now let us keep our fast."

II

How it is better not to fast than to boast about our fasting - as the Lord saith, "When ye fast, appear not unto men to fast."

There was an assembly of monks in a certain church on a feast day. As the custom was, after the sacrifice had been offered among them, the brethren dined together. One of them said to the disciple who set food before him, "I will not eat this. I eat no cooked food." This he said boasting of his own abstinence. Then said the blessed Theodorus, "It would be better for you, brother, to be eating flesh in your own cell, than that such a word should be heard among the brethren."

III

How humility is to be preferred before fasting. A certain anchorite dwelt in a cave not far from a monastery, and led a life of great privation. Once some brethren came from the monastery to visit him. As the custom was, he set food before them to refresh them after their journey. The brethren compelled the old man to eat with them, saying that they would not eat without his company. Afterwards, when they thought upon what they had done, they said to him," We fear that you are grieved, father, because to-day for our sakes you have eaten more than you are wont." But he replied, "Brethren, I am not troubled in this matter. I am only grieved when I have acted according to my own will."

IV

How charity is to be preferred to fasting.

Epiphanius, the Bishop of Cyprus, once sent a message to the holy Hilarion, saying, "Come hither, that we may see each other and converse together before we depart from the body." Hilarion came, and the two old men sat down to eat together. There was set before them the flesh of some birds. Of this the Bishop partook, but Hilarion refused it, saying, "Pardon me, but since I became a monk I have never eaten anything that had life." At these words the Bishop was grieved, and replied, "Since I became a monk I have tried never to allow anyone to sleep until I had removed any cause of complaint he had against me, nor myself to go to sleep while I was vexed with anyone." "My father," said Hilarion, "I pray you pardon me. Your way of life is far more excellent than mine."

V

The saying of an unknown monk, teaching the same thing.

It is better to eat meat and to drink wine than to feed upon the flesh of your brother by envying him.

VI

The teaching of St. Antony, that wisdom is to be preferred to fasting.

There are some who keep under their bodies by fasting, and yet are far from God because they lack discretion.

VII

The teaching of the abbot Moses on fasting as an aid to perfection._

Fastings, vigils, meditations on the Scriptures, self-denial, and the abnegation of all possessions are not perfection in themselves, but aids to perfection. The end of the science of holiness does not lie in these practices, but by means of them we arrive at the end. He will practice these exercises to no purpose who is contented with these as if they were the highest good. A man must not fix his heart simply on these, but must extend his efforts towards the attainment of his end. It is for the sake of the end that these things should be cultivated. It is a vain thing for a man to possess the implements of an art and to be ignorant of its purpose, for in it is all that is of any value.

VIII

The teaching of the abbot Theonas about the occasions on which men ought not to fast.

If at the coming of a brother, in whose person a man ought to refresh Christ with courtesy and embrace Him with a kindly welcome, he should choose to observe a strict fast, would he not be guilty of churlishness rather than be deserving of praise for devoutness? If, when the failure or weakness of the flesh requires the strength to be restored by partaking of food, a man will not consent to relax the rigour of his fasting, is he not to be regarded as a cruel murderer of his own body rather than as one who is careful for his own salvation? So, too, when a festival season permits a suitable indulgence in food and a liberal repast, if a man will resolutely cling to the strict observance of his fast he must be considered as not religious, but rather boorish and unreasonable.

IX

How spiritual thoughts put to silence the demands of the body.

Once there came a hermit to the cell of an elder to talk with him. The elder said to his disciple, "Prepare some vegetables for us, and moisten some bread." The disciple did so. But the two old men remained in spiritual converse till the sixth hour of the next day.

Then said the host again to his disciple, "Prepare some food for us." The disciple answered him, "My father, I prepared it yesterday." Then the two old men rose up and ate together.

X

Of a certain brother who conquered his body lest he should grieve another.

One of the elders was sick, and for many days could not eat. At last his disciple asked to be allowed to prepare a special dish that he might relish. Now there was in the cell a jar in which there was a little honey. Beside it there hung another containing oil, and that rancid, for the lamp. The disciple by mistake poured the oil and not the honey on the dish he had prepared. The old man, when he had tasted it, said not a word but silently swallowed a mouthful. The disciple then constrained him to eat some more. With difficulty he did so. Again the disciple pressed him to take of it a third time. But the old man replied, "In truth, I cannot eat again, my son." The disciple still pressed him, saying, "It is very good. See, I will eat with you." When he tasted the dish, and knew what he had done, he fell upon his face and said, "Alas, my father, I have poisoned you. Why did you not speak?" Then the old man said, "Be not grieved, my son. If it had been God's will for me to eat honey then you would have put honey in your dish."

XI

The use of fasting, and how it helps the life of the soul.

Fasting is the bridle in the mouth of the monk. It holds him back from sin. He who rejects the practice of fasting is like an unbridled, fiery horse. He is swept away by passion.

XII

The conduct of the abbot Moses, and how the brethren recognized that charity is above rubrics.

Once a rule was made in the Scetic desert that the monks should fast during the week of the Passover. It happened, however, that certain brethren from Egypt came to visit the abbot Moses during that very week, and he prepared some food for them. Some of the neighbouring monks saw the smoke of his fire rising from Moses' cell, and they said to the clergy of the church which was there, "Lo! Moses has broken our rule and cooked some food." Then the clergy replied, "When he comes we will speak to him about the matter." On the Sabbath, when the abbot Moses came with the strangers to the church, the clergy understood his conduct, and cried out in the presence of the assembled brethren, "Oh, abbot Moses, you have indeed broken a commandment of men, but you have bravely kept the commandments of God."

XIII

A rule of life.

A certain brother once visited a hermit, and was entertained by him. He feared lest his entertainment had interfered with the severity of the hermit's living, and when he was departing he said, "My father, pardon me if I have hindered the observance of your rule of life." The hermit answered him, "My rule of life is to receive you with hospitality, and let you depart in peace."

XIV

How a man may break his fast through love, and another who keeps his fast may yet be yielding to a base kind of self-indulgence.

Once there were some brethren who, for the love they bore their guests, ate with them, though it was a season of fasting. There was another brother who scorned them as they sat at meat. When the abbot John beheld him he wept, saying, "What kind of spirit has this man in his heart that he laughs at the brethren, scorning them? He ought rather to be weeping for himself. It is he who breaks his fast, not they. It is he who is eating. He devours charity."

XV

It is better not to fast than to be praised for fasting.

In a certain region there was a man who fasted much, so that the name of Faster was given to him. Hearing this the abbot Zeno sent for him. He came joyfully. After praying together they sat down, and the abbot Zeno began to work in silence. Having no chance of speaking, the Faster was attacked by a restless spirit of accidie. At last he said, "Pray for me, my father, for I am going away." "Why are you going ? " asked the old man. " Because," said the other, "my heart is as if it were on fire, and I know not what is the matter. When I was at home I used to fast until the evening time, and no such thing happened to me." Then said the old man, "At home you were fed through your ears by men's praises. Now, go away. Eat at the ninth hour, and if you do anything, do it secretly." In following this advice he found that he came to look forward eagerly to the ninth hour. Those who knew him began to say of him, "The Faster has fallen under the power of some devil." He then came and told all this to the abbot Zeno, who said to him, "This way and this leading is according to God's will."

CHAPTER 10

On Poverty

If thou wilt be perfect, go and sell that thou hast, and give to the poor . . . and come and follow Me.

- *St. Matt.* xix. 21.

Sell that ye have, and give alms; provide yourselves bags which wax not old, a treasure in the heavens that faileth not, where no thief approacheth, neither moth corrupteth.

- *St. Luke* xii. 33. If thou wilt enter into life, keep the commandments. If thou wilt know the truth, believe Me. If thou wilt be perfect, sell all. If thou wilt be My disciple, deny thyself utterly. If thou wilt possess a blessed life, despise this life present.

- *The Imitation of Christ*, iii. 56. Keep this short and perfect word: Let go all, and thou shalt find all; leave desire, and thou shalt find rest.

- *The Imitation of Christ*, iii. 32.

VOLUNTARY poverty is half-way between the kind of asceticism which we have called physical and that which may properly be described as spiritual. On the one hand, it is clear that poverty like that of the hermits deprives a man not only of all the luxuries of life, but of what are generally regarded as its necessary comforts. On the other hand, the sin which stood in direct antithesis to their conception of poverty was covetousness; and this is a sin of the soul, not of the body.

The absolute renunciation of all property was the initial act of the hermit's entrance upon his new life. From the point of view of the fathers of monasticism, the necessity for this renunciation was obvious. Every possession was a tie to the world, and the great object was to get free of the world, to stand clear of its ambitions, its pleasures, and its cares. A man who possesses property, even if he is content to forego the possibility of increasing it, must yet take care to preserve it. He must dedicate some portion of his time, his ability, and his energy to the getting or the management of his income. All such care and expenditure of strength was, from the hermits' point of view, a service of mammon, and they remembered the Lord's words - "Ye cannot serve God and mammon." There was no point, therefore, of their life on which the hermits insisted more vigorously than the completeness of the original renunciation. What the postulant ought to do with his money was not definitely settled. Sometimes it was given to his relatives, sometimes it was handed over to the clergy for the use of the church. Oftenest, perhaps, in strict obedience to the Lord's command, it was given to the poor. Whatever the destiny of the money might be, it was essential for the hermit to aet rid of it entirely. No half measures were tolerated. The parable which St. Antony made the young monk act, who wanted to keep something for himself, is almost savage in the intensity of its insistence on absolute renunciation. The personal possessions which a monk might retain were not, any more than the manner of his fasting, settled by definite rule. That their theory of poverty was spiritual, as opposed to mechanical, may be seen in the saying which described true poverty as the possession of nothing which it would cost a pang to give away. He who lives in such poverty as this places no obstacle in the way of his fulfillment of the Lord's words - " Give to him that asketh thee." How complete the renunciation occasionally became may be seen in a fine story of Besarion. He owned nothing in the world but a cloak, an undergarment, and a copy of the gospels. Once, as he went upon a journey, he threw his cloak over a dead body which lay exposed on the roadside. Further on his way he gave his other garment to a naked beggar. Then, moved by the recollection of the Lord's words, he sold his copy of the gospels and gave the proceeds to the poor.

Even, however, when the initial act of renunciation was as complete as possible, there still remained for the hermit the possibility of being ensnared by covetousness or entangled in worldly cares. It must not be forgotten that the hermits were diligent workers. They preferred such kinds of work as could be done in or near their cells. They wove mats and baskets, or cultivated little gardens; the fruits of their labor they sold, sometimes carrying them to neighboring villages, sometimes sending them in boat-loads down the Nile to the great cities. At harvest-time they frequently hired themselves out as laborers. The money thus earned they used first for the supply of the few necessities of their own lives, and what remained for the relief of the poor. The marketing of their goods was, as may readily be supposed, a distasteful task. Haggling and bargaining involved them in what must always be a degrading struggle. Some of them simply named a price for their goods, and then, if they were offered less, took it without protest. Others declined even to name a price. They exposed their wares in the market-place, and took the price offered by the first buyer who approached them.

Even, however, when their traffic was regulated by these principles, there remained a possibility of covetousness. There are grievous stories of men who hoarded little stores of money. Sometimes the motive seems to have been mere desire of possession. Sometimes it was, at first at all events, a less unworthy one. It was in order to make some provision for future sickness that the brother, whom the angel healed, began to lay by some portion of his earnings. All such saving was regarded as displaying, at the least, a lamentable want of faith. The ideal of the hermits was a perfect trust in Him who feeds the ravens and clothes the lilies of the field. To save and make provision for the future was to call down the Lord's rebuke - " Oh, ye of little faith."

I

How a certain brother understood the words of the Lord very literally.

A certain old man was once asked by one of the brethren what a monk ought to do to be saved. The old man took his raiment and stripped it off. Then, stretching forth his hands, he said, "Thus ought a monk to be naked of all that belongs to this world. Thus also should he stretch himself out in crucifixion, that he may come out conqueror from the

68

temptations and struggles of this world."

II

The advice of St. Antony to a disciple who desired to be a monk, and yet was unwilling to give away all that he had.

A certain brother renounced the world, and gave what he possessed to the poor. Yet, because he was fearful of heart, and had little faith, he retained somewhat in his own power. This man paid a visit to St. Antony. When the saint perceived how the case was with him, he said to him, "Go thou to yonder village. There buy meat, and bind it with cords round thy naked limbs. Then return to me." The disciple did so, and lo! as he was returning to the saint the dogs from the village and afterwards the birds of the air, tore his limbs, grasping at the meat bound to them. On his return, the saint asked him how he had fared, he replied by displaying his wounds and blood. Then said St. Antony, "They who renounce the world, and yet desire to possess money, lo! like dogs and birds, the demons strive with them and tear them."

III

Of the measure of renunciation, and when it may be regarded as complete.

An old man said, "Own nothing which it would grieve you to give to another, nothing which would lead you to transgress the commandment of the Lord - 'Give to him that asketh you.'"

IV

The word of Serapion to a monk who owned what he was unwilling to part with.

A brother asked the abbot Serapion to speak some word of exhortation to him. Serapion said, "What can I say to you, seeing that you have taken the property of the widow and the orphan and put it on the window-sill of your cell?" He said this, having seen that this brother had many books which he kept in his window.

V

How the same Serapion who spoke thus had himself made a perfect renunciation.

One of the monks, a certain Serapion, possessed a copy of the gospels. This he sold, and gave the price of it to the poor and hungry. Then he went home rejoicing, saying to himself, "Lo! now I have sold even that very book which was for ever saying to me, 'Sell all that thou hast and give to the poor.'"

VI

A description of the sin of covetousness, through which men fail in making their renunciation perfect.

We must not only guard against the possession of money, but also expel from our souls the desire of possessing it. For it is necessary not so much to avoid the results of covetousness, as to cut off by the roots all disposition towards it. It will do no real good not to possess money, if there exists in us the desire of getting it.

VI

The story of a monk who fell before a very subtle temptation, but in the end was saved.

The elders relate a story of a certain monk who was a skilful gardener. He laboured diligently, and all that he earned he gave to the poor after he had supplied his own necessities. After a while Satan found entrance into his heart, and said to him, "Keep something of what you earn for yourself. Some day you will be old or fall sick, and then you will have need of what you can save now." It seemed wise to the monk to do this, and he saved until he had filled a large pitcher with coins. It happened that he fell sick, and an abscess gathered on his foot. He expended all that he had saved on doctors, neither was made any better. At last one of the most skilful doctors said to him, "Unless your foot is cut off you cannot recover. And they fixed a day for the amputation of his foot. That night he came again to his right mind, and wept bitterly for what he had done, being truly repentant. Then, groaning frequently, he prayed, and said, "Be mindful, O Lord, of the work which once I did, how I labored in my garden and gave the reward of my labor to the poor." When he had so prayed, behold an angel of the Lord stood by him and spoke to him, saying, "Where is now the money you saved? Where is the hope with which you saved it?" He, understanding well what the angel said, replied, "I have sinned! O Lord, pardon me. Henceforth I will do no such things as these for which you reproach me." Then the angel touched his foot, and immediately it was healed. In the morning he arose and went forth to labour in his garden.

VIII

How all we give, we give to God, and not to men.

Melania relates that she brought three hundred pounds of silver to the abbot Pambo, and asked him to accept the gift for the use of the monks who were in need. He said to her, "May God give you your reward." Then, turning to his servant Theodore, he said, "Take this money and distribute it among the brethren who dwell in Libya and in the islands, for the monasteries there are very poor." Melania, in the meanwhile, stood waiting for his benediction, and expected that he would speak some word of praise to her for the greatness of her gift. At length, when he remained silent, she said, "Master, do you know how much I have given? There are three hundred pounds of silver." But Pambo took no notice of her, and did not even glance at the boxes of money. At length he replied, "He to

70

whom you make this gift, my daughter, does not need that you should tell Him how much it is. If you were giving this money to me, you would be right to tell me the sum of it. Since, however, you are giving this money to God, who did not despise even the two mites, but valued them above all other gifts, you may well be silent about the amount of it."

IX

How a hermit refused to receive a gift of money, even for the use of the poor.

A certain man asked a hermit to receive a gift of money for his own use. He refused, saying that the earnings of his labor sufficed him. The other, however, besought him to take the money and use it for the poor, if not for himself. The hermit replied, "So I should run a double risk. I should take what I do not want. I should distribute what another gave, and be praised."

X

You cannot serve God and Mammon.

A certain brother once came to an elder, and said, "My father, of your kindness tell me, I beseech you, what I ought to strive for in my youth, that I may own something in my old age." The old man replied to him, "You may either gain Christ or gain money. It is for you to choose whether you will have for your God the Lord or mammon."

XI

The story of three monks who were not greedy for money.

Once three brothers hired themselves out as harvest laborers, and agreed together to reap a certain field. On the very first day of their labor one of them fell sick and returned to his cell. The other two remained, and one of them said to the other, "You see how a sickness has fallen upon our brother so that he cannot work. Do you therefore do violence to yourself, and I shall do likewise. We shall put our trust in God. Our brother who is sick will pray for us. It may be that we shall be enabled to do double work and reap his part of the field as well as our own." They did as they had hoped, and reaped the whole field which they had undertaken. On their way to receive their wages they called the brother who was sick, saying, "Come, brother, and receive your pay." But he said, "What pay shall I take, seeing I did not reap." They replied, "It was through your prayers that the reaping was accomplished. Come, therefore, as we say, and get your wages." Then there was strife between them, for he kept saying, "I will take no pay, for I have done no work"; and they refused to take any wages at all unless he got his share. At last they referred the matter to the judgment of a certain renowned elder. The brother who had been sick told his story first: "We three went to reap a certain field for hire. When we came to the place where we were to work, on the very first day I fell sick. I returned to my cell, and from that time on I did no work at all. Now these brethren come to me insisting and saying, 'Brother,

71

come, take pay for work you did not do!'" Then the other two brethren spoke and said, "We did, as he says, go to work, and did undertake to reap a certain field. It was such a field that if we had all three been there we could hardly by great toil have fulfilled our task. Yet through the power of this brother's prayers we two were able to reap the whole field more quickly than the three of us expected to do it. Now when we say to him, 'Come and receive your hire,' he will not do so." When the old man who judged between them heard their stories, he marveled greatly. Then he said, "Give the signal for the brothers to assemble." When they were gathered together he said, "Listen, brethren, to the righteous judgment which I give." Then he told them the whole story, and gave his decision that the brother who had been sick should receive for his own the share of the pay which ought to have been his. That brother, however, departed sorrowful, like one to whom an injury is done.

CHAPTER 11

On Obedience

I came not to do Mine own will.

- *St. John* vi. 38.

Be desirous, my son, to do the will of another rather than thine own.

- *Imitation of Christ*, iii. 23.

Thirty years of our Lord's life are hidden in these words, "He was subject unto them."

- *Bossuet.*

OBEDIENCE is the sacrifice of self-will. It may consist passively in a man's refusing to insist on acting in accordance with his own conception of what is pleasant, or his conviction of what is expedient or right. It may involve an act or a course of action directly opposed to such convictions. The Egyptian hermits recognized unquestioning obedience as a great virtue. The language in which they praise it is fervid. Its place in the hierarchy of virtues is supreme. The examples which are quoted for imitation show that no idea of compromise was to be entertained. It is apparent at once that the general conscience of mankind endorses under certain circumstances all that the hermits taught about obedience. The citizen of a state must submit to the will of the power that governs. The soldier must obey promptly and unquestioningly the orders of his officer. The sailor has no right of self-assertion against the will of his captain. No consideration of the justice or injustice of a law will absolve a citizen from obeying it so long as it continues to be the law. No conviction of the folly or inexpedience of an order can be held to justify the mutiny of the soldier or the sailor. Under certain circumstances we are as much convinced as the hermits were that obedience without delay or protest is an essential duty - is even the highest virtue. Of all conceivable circumstances only one is generally held to justify disobedience. If obedience involves, directly and unmistakably, a transgression of the law of God, then every man, citizen, soldier, sailor, or monk is held to be right in disobeying.

So far there is nothing in the hermits' position about obedience which seems to conflict with the feeling even of men fundamentally opposed to monasticism. Nevertheless, there is felt to be a difference somewhere. A man who willingly recognises the soldier's obedience to his officer as a virtue, finds a feeling of irritated hostility arise in his mind at the contemplation of a monk's obedience to his abbot. Here, as very often elsewhere, a feeling which is, as one may say, instinctive to many men, is found upon examination to have a reasonable justification. The obedience of the hermit is a different thing from that of the soldier or the sailor. It rests upon a different basis, aims at a different result. The soldier obeys because, without discipline, an army is a useless mob. The sailor obeys because considerations for the common safety necessitate the predominance of one man's will. If the conditions which necessitate obedience are removed. obedience itself ceases to be a virtue, and may become even a vice. When a volunteer regiment is disbanded, at the end of a war, the trooper no longer owes, or is supposed to owe, any kind of obedience whatever to the man who was his officer. When a ship comes to a port, and the crew is paid off, the sailor has no special duty to his captain. This is only to say that obedience is regarded simply as a necessary condition for success in all cases where combined effort is required. Once the success is attained there is no more reason for obedience. Apart from the obvious advantages of discipline, obedience - that is, obedience simply for the sake of obeying - strikes the ordinary conscience as silly, if not actually wrong. The hermits looked at the matter altogether differently. To them obedience was not a means of perfecting any organization, but was a virtue in itself. It was one of the marks of the ideally perfect character. A hermit obeyed his abbot or his elder brother because he wanted to be good, and being good involved the total conquest of that self whose outworks were passions and lusts, but whose last stronghold was the desire to express in act its own convictions and will. Here we see why in the case of the hermit the wisdom or folly, the expediency or inexpediency of the command given were quite unimportant. John of Lycopohis was ordered, when he was young, to plant and water a dried-up stick. In itself the command was a silly one. Neither planting nor watering made any difference to the stick. Obedience or disobedience did, however, make all the difference possible for John. He obeyed, and by obeying built up within himself a certain character. He so far annihilated self and self-will that it ultimately became possible for him to receive direct revelations of the divine purpose. He might have disobeyed. Then also he would have built up a character - forceful, dominant, masterful - but not such as enables a man to be the intimate friend of Jesus Christ.

The judgment which condemns obedience like John's as a worthless waste of time and energy is based upon a mistaken estimate of the relative value of what a man is and what he does. John, and others like him, might have spent their time in doing things that would have seemed to us more useful. Supposing that they had, the value of their work would be all exhausted centuries ago. The fields they dug would have gone back to barrenness or been dug again a thousand times. But the character which these men built up, by God's grace, is to-day, as we believe, in Paradise a joy to the angels, a glory to the Master whom they served. By asserting themselves against a command which seemed foolish they might have accomplished something effective for a year or two. They might have cast deeds, like stones, into the pool of human life, have watched the waters splash and ripple, and close calm again. By obeying they built into eternity, reared the fabric of a beautiful and everlasting human soul.

I

The praise of the virtue of obedience.

Oh, my son, good indeed is that obedience which is rendered for the Lord's sake. See to it, therefore, that your feet are placed upon the pathway that leads to the perfection of obedience. In obedience is the safety of all faithful souls. Obedience is the mother of all kinds of virtue. Obedience discovers the road that leads to heaven. It is obedience that opens heaven's gates and raises men above the earth. Obedience hath her home among the angels. Obedience is the food of all the saints. From her breasts they suck the milk of life, and grow up to the measure of perfection.

II

The vision which one of the fathers saw, wherein was manifested the greatness of obedience.

One of the fathers, being in a trance, saw four kinds of men standing before God. First he saw those who, though they suffer in the body and are sick, yet give thanks to God. Next were those who give themselves to hospitality and are devoted to the relief of others' needs. Next were they who dwell in solitude and see not the faces of men. The fourth kind were they who strive to be obedient and submit themselves unto the will of the fathers. He beheld, and lo! this last kind was superior to the other three. They were wearing golden crowns, and had received an excellent glory above the glory of the rest. Then the old man spoke to him who showed the vision to him, saying, "Why has this fourth kind of men a greater glory than the others?" He was answered thus, "They all find some satisfaction in doing the things they wish to do, albeit the things they wish are all of them good. He, however, who obeys renounces the fulfillment of his own will. He gives himself up to the will of the father who orders him. Therefore to his share there falls an excellent glory above the glory of the rest."

III

How obedience is no virtue if we only render it to those whose commands are according to our inclinations.

There was a brother once who said to a famous elder, "Father, I wish to find an old man with whom to dwell. I seek for one whose ways will be altogether according to my ideas of what is right. With him I wish to live and die." The elder said to him, "Of a truth you are on a noble quest, my master." Then he repeated his desire, being proud of it, and not understanding the meaning of the other's words. When the elder perceived that he still regarded his desire as a good and noble one, he said to him, "Then, if you find an old man whose ways answer to what you think is right, do you think that you will stay with him?" The brother replied, "Certainly I should stay with such a one if he indeed answered to my expectations." Then said the elder again, "Do you not see that you would not be following the teaching of him whom you seek for a master? You would be simply walking according to your own will." Then that brother, understanding what the old man said to him, fell at his feet in penitence and said, "Pardon me, for certainly I have boasted greatly. I thought that I was saying what was good, and all the while there was no trace even of goodness in my words."

The obedience of Mark, the disciple of Silvanns.

The abbot Silvanus had a disciple whose name was Mark. He was remarkable for his obedience, and therefore Silvanus loved him. Now he had also eleven other disciples, and they were vexed because Mark was more beloved than they were. When the elders heard this they were grieved, and came to Silvanus, intending to ask him to give up his favorite, since the brethren were offended. Before they had said anything Silvanus took them with him to make a round of the various cells. He called each monk by his name, saying, "Come out, for I have work for you to do." No single one of them was willing to come out. The last of all of those to whom they came was Mark. Silvanus knocked at his door, and called his name. Immediately Mark came out, hearing his master's voice. Then the abbots entered Mark's cell. Now Mark was a writer who copied books. Looking at the manuscript at which he had been at work, they found that he had left unfinished the letter which he was forming when he heard the voice of the old man. This he had done that his obedience might be prompt. Then said the other elders to Silvanus, "Truly him, whom you love, we also love; and no doubt God loves him because of his obedience."

CHAPTER 12

On Avoiding the Praise of Men

Take heed that ye do not your alms before men, to be seen of them.

- *St. Matt.* vi. 1.

When thou prayest, thou shalt not be as the hypocrites are: for they love to pray . . . that they may be seen of men.

- *St. Matt.* vi. 5.

When thou fastest, anoint thine head, and wash thy face that thou appear not unto men to fast.

- *St. Matt.* vi. 17, 18.

Never desire to be singularly commended or beloved, for that pertaineth only unto God, who hath none like unto Himself.

The Imitation of Christ, ii. 8.

Mere empty glory is in truth an evil pest, the greatest of vanities; because it draweth man from the true glory, and robbeth him of heavenly grace.

The Imitation of Christ, iii. 40.

THAT is a fine saying in which vainglory is compared to an onion or other bulbous root. In the region of spiritual asceticism there is no struggle more difficult than that against the spirit of vainglory. The desire of being praised - and this is what the hermits meant by vainglory - is natural to every man, Christian or pagan, good or bad. In whatever sphere of human activity a man may elect to spend his energies, the praise of some men will wait for him. One man may desire and work for the praise of the crowd, another may find a subtler measure in the congratulations of the few. To one it is enough that the multitude should reckon him to be a good man and throng to listen to his teaching. To another the recognition of his merits by the multitude seems in itself a kind of condemnation. He desires the less audible approbation of the one or two whose own righteousness constitutes them fit judges of what is good. Some men are found openly exulting in being praised. No flattery is too coarse or obvious for them. When it is withheld they demand it blatantly. Others shrink from the sound of open praise, and yet go through life, cautiously feeling about for signs of the esteem in which their neighbors hold them. The hermit who compared the love of praise to an onion had probed far down into human weakness. His sight was keen when he saw that to escape the desire of praise for one kind of virtue is to find oneself seeking it all the more earnestly for another, until the soul is caught in the paradox of desiring to be known as one who does not wish for praise at all.

Vainglory must not be confused with pride. It is the strong man who is proud. In proportion as he grows stronger he feels less and less need for the approbation of others. Milton's heroic Satan may stand as a type of strength and pride. We do not think of him as troubled much about any judgment passed on him. He neither seeks praise nor dreads blame. It is our weakness which makes us long for approbation. We are not sure enough of ourselves to stand alone or persevere without someone to tell us we are doing well. Thus pride and vainglory are opposed to each other. They are the besetting sins of opposite types of character. A man may even be cured of overmuch desire of praise by teaching him to be proud enough to disregard the opinions of the crowd about his acts. Yet it was not because vainglory was an indication of weakness that the hermits strove so hard against it, nor was it along the way of pride and strength that they sought to escape. They thought of virtue as such a tender plant that the breath of praise withered it. Goodness, in their opinion, actually ceased to be the highest kind of goodness when it was recognized. The ideal was to live and die unknown. I do not remember that the hermits ever appealed directly to the example of the Lord in their shrinking from vainglory, but I am sure that their teaching was entirely in accordance with the spirit of His life. For far the greater part of the time of His dwelling among us He chose to remain unknown. Even when the fulfillment of His mission involved His doing works which some men were sure to praise, He strove by all means to avoid publicity. The very manner of His great sacrifice of Himself was so devoid of all obvious heroism that it was only after its consummation that His lifting up began to draw all men unto Him.

Just as it was not because the desire of praise is a sign of weakness that the hermits condemned it, so it was not by trying to be strong and independent that they avoided it. The story of the abbot Nisteros' flight from the serpent is so quaint that at first the reader is moved only to smile. Yet in it we find a man avoiding the peril of being praised by a display of weakness and even cowardice. So, too, the abbot Sisois does not try to attain that position of haughty isolation which would have made him indifferent to the judge's praise or blame. He, like Nisteros, in order to avoid vainglory, deliberately courts contempt. He aims at being despised, lest the Lord's "woe" should fall upon him, and men learn to speak well of him.

I

A saying concerning virtue, how it should be hidden.

A certain one said, "As treasure when it is discovered speedily becomes less, so virtue made known unto man vanishes. As wax melteth at the fire, so the virtue of the soul is thawed and runs away when it is praised."

II

A warning against the danger of being praised.

A brother once asked the abbot Mathoes: "If I go to dwell in any place, what shall I do there?" The old man answered him, "If you dwell in any place, do not make a name for yourself there for anything. Do not say that you will not join the meetings of the brethren, or that you will not eat this or that. So doing, you will make a name for yourself. Afterwards you will perhaps be praised and become famous. Then others will come to inquire of you concerning the way of life, and your own soul will be injured by their frequent comings."

III

"Love to be unknown."

The abbot Zeno, the disciple of Silvanus, said, "Never dwell in a famous place, or make a friend of a famous man."

IV

The advice of the abbot Macarius to those who desire eminence.

St. Macarius once said, "Do not desire, nor, if you can help it, permit yourself to be made the head of a congregation, lest perhaps you lay the weight of other men's sins upon your neck."

V

A story of the abbot Nisteros, how he escaped the temptation of vainglory.

The abbot Nisteros the elder was one day walking in the desert with one of his disciples. Seeing a serpent in their path, they both turned and fled from it. Then the disciple said, "My father, were you afraid?" The old man answered him, "I was not afraid, my son, but it was better for me that I should flee before the serpent. If I had not at once fled from it, I should afterwards have had to flee before the spirit of vainglory."

VI

A story of the abbot Sisois, how he avoided being praised by one who wished to admire his way of life.

On one occasion a certain judge wished to pay a visit to the abbot Sisois. Some of the clergy went beforehand, and said to him, "Father, prepare yourself, for the judge has heard of your works and your piety, and is coming to visit you. He desires also to receive your benediction." Sisois said, "I shall do as you desire. I shall prepare myself for his visit." Then he clad himself in his best garments, took bread and cheese in his hands, and seating himself with outstretched feet at the door of his cell, began to eat. When the judge with his retinue arrived and saw him, he said, "Is this the famous anchorite of whom I heard so much?" So, despising Sisois, he departed.

VII

A comparison which shows the nature of vainglory.

The elders admirably describe the nature of this malady as like that of an onion, and of those bulbs which when stripped of one covering you find to be sheathed in another; and as often as you strip them you find them still protected. All other vices when overcome grow feeble, and when beaten are rendered day by day weaker. But vainglory, which is the desire of praise, when it is beaten rises again keener than ever for the struggle. When we think it is destroyed it revives again, and is stronger than ever on account of its death. The other kinds of vices only attack those whom they have overcome in the conflict. This one pursues those who are victorious over it all the more keenly. The more thoroughly it has been resisted, so much the more vigorously does it attack the man who is elated by his victory over it.

VIII

A word of St. Antony, teaching that he who suffers himself to be counted foolish, alone is wise.

Some of the elders once visited St. Antony, and with them came the abbot Joseph. St. Antony, wishing to prove what manner of men they were, started a question about the meaning of a passage of Scripture. One by one they gave their opinions about the meaning of it. To each of them he said, "You have not hit it." At last it came to the turn of the abbot Joseph, and the saint said to him, "In what way do you understand this passage?" He replied, "I do not know." Then said St. Antony, "Truly, the abbot Joseph has discovered the way in which Scripture is to be interpreted, for he acknowledges his own ignorance."

IX

Of the subtlety of the temptation of vainglory, which is the pleasure of being praised by men.

Our other faults and passions are simpler, and have each of them but one form. This one takes many forms and shapes, and changes about and assails the man who stands up against it from every quarter, and assaults even him who conquers it on every side. It tries to find occasion for injuring the servant of Christ in his dress, in his manner, his walk, his voice, his work, his vigils, his fasts, his prayers. It lies in wait for him when he withdraws to solitude, when he reads, in his knowledge, his silence, his obedience, his humility, his patience. It is like some most dangerous rock hidden by the waves. It causes miserable shipwreck to those who are sailing with a fair breeze, while they are not on the look out or guarding against it.

X

A rebuke of ostentation.

There was a certain brother who practiced abstinence from various kinds of food, and especially refused to eat bread. He went once to visit a renowned elder. As it happened, while he was there, some strangers arrived, and the old man prepared a scanty meal for them. When they sat down to eat the brother who practised abstinence would eat nothing except a single bean. When they rose from the table the elder took him apart privately, and said to him, "Brother, when you are in the company of others do not be anxious to display your own way of living. If you really wish to keep your rule of life unbroken, sit in your own cell and never leave it." When he heard these words he felt that the elder was right. Therefore ever afterwards he conformed his ways to those of the brethren among whom he found himself.

CHAPTER 13

On Anger

Every one who is angry with his brother shall be in danger of the judgment; and whosoever shall say to his brother, Raca, shall be in danger of the council; and whosoever shall say, Thou fool, shall be in danger of the hell of fire.

- *St. Matt.* v. 22 (R.V.).

Nothing so stills the elephant when enraged as the sight of a lamb; nor does anything break the force of a cannon-ball so well as wool. Correction given in anger, however tempered by reason, never has so much effect as that which is given altogether without anger; for the reasonable soul, being naturally subject to reason, it is a tyranny which subjects it to passion, and whereinsoever reason is led by passion it becomes odious, and its just rule obnoxious.

- *St. Francis of Sales, The Devout Life*, viii.

THE only point which is really peculiar in the hermits' teaching about anger is that the possibility of righteous anger is altogether denied. No matter how wicked a brother might be, or how serious the consequence of his sin, it was not right to be angry with him. To try to cure another of sin by angry denunciation was the same thing as for a physician to try to cure his patient by innoculating himself with a similar fever, for to be angry even with sinfulness is to sin.

Apart from this one point, the hermits' teaching is only remarkable for the accuracy of its analysis of the source from which anger springs, and its thoroughness in the practical treatment of the fault.

Anger is traced back to the hermits' most intimate enemy - self. It is an expression of selfishness, a sign that self has not wholly and really been conquered. Thus anger may spring from avarice. It is then the protest of self against any interference with what are regarded as possessions. Where the renunciation of property is really complete this kind of anger becomes impossible. There is a beautiful story of two hermits who determined to find out by experience what it was like to be angry. They planned that each of them should claim for his own an earthen pot which lay in their cell. The attempted quarrel began well enough, for the first monk said, "The pot is mine," and the second replied to him, "No, it is mine." But at this point the first man's resolution broke down, and he said, "As you say, brother, it is yours." This hermit had so entirely renounced the satisfaction of possessing anything that it was as impossible for him to grow angry in a dispute about property as it would be for a sensible man to do battle with a child for the sake of some treasure of broken glass or colored stone. The desire of impressing his own will or opinions upon others is another sign that the old self in a man is not wholly dead. Where such a desire exists in any strength, and others thwart it, the result is anger. In the same way vainglory, when it is starved for want of praise, and pride when it proves to be indulged in foolishly, give birth to anger. Vainglory and pride are alike vices of selfishness.

The hermits distinguished various stages of anger, to each of which was attached a certain degree of guilt. There was first the feeling of anger in the heart, the sudden rush of bitter feeling consequent on suffering unjustly. This cannot be fought against. It may be avoided only by those in whom the old self is utterly dead. Next comes the expression of anger on the countenance. It is at this point that the hermits' battle with anger really begins. It is possible to choke down at once the emotion so that not even the tightened lips or frowning brow betray its presence. Then there is the vent which anger finds in words. Here is another point of defense for the hermit. He may and ought to be able to bridle his tongue. The final stage of anger is when a man so loses self-control as to strike or injure another. It is something to have stopped short of this.

There is an altogether different kind of anger, which has its origin not in the negative side of the religious life, through failure to eradicate the old selfish instincts, but in the positive side, in coming short of absorbing interest in divine things. To the hermit who fell away from his loving desire for the Lord, whose mind ceased to be dominated by visions of the King in His beauty, the life of the cell or the community became an intolerable weariness. A craving for change and excitement seized upon him. The monotony of his daily round alternately oppressed and goaded him. In this condition he was a ready prey to peevishness and irritability. He flew into sudden fits of unreasoning fury with brothers who had in no way offended him; or if human objects were absent, vented his ill-humor by cursing his pen or his knife or the stones on the road when his feet tripped on them. This kind of anger was the result of a morbid spiritual state which the hermits recognized as sinful, and called accidie. To fly from the circumstances which gave excuse for its expression was manifestly useless. It is possible to fly from men but not, as the hermit in the story found, from the demon who excites to this kind of anger. Even the attainment of a sleepy apathy is not a real cure for it. The serpent is venomous Still, though he lies torpid and bites no one. The true cure lies in the renewal of the broken communion with God. Then the weariness and accidie give place to active joy, and the temptation to sudden anger-fits disappears.

I

The teaching of a certain elder, concerning the nature and origin of anger.

A certain elder said, Anger arises through four things - through the greed of avarice, whether in giving or receiving; also through loving and defending one's own opinion; through a desire of being honorably exalted; also through wishing to be learned or hoping to be wise above all others.

In four ways anger darkens the nature of a man - when he hates his neighbor, when he envies him, when he despises him, and when he belittles him.

In four places anger finds scope - first in the heart, second in the face, third in the tongue, fourth in the act. Thus if a man can bear injury, so that the bitterness of it does not enter into his heart, then anger will not appear in his face. If, however, it find expression in his face, he still may guard his tongue so as to give no utterance of it. If even here he fail and give it utterance with his tongue, yet let him not translate his words into acts, but hastily dismiss them from his memory.

Men are of three kinds, according to the place which anger finds in them. He who is hurt and injured, and yet spares his persecutor, is a man after the pattern of Christ. He who is neither hurt himself, nor desires to hurt another, is a man after the pattern of Adam. He who hurts or slanders another is a man after the pattern of the Devil.

II

How we must not suppose that the spirit of anger is dead in us when we happen to escape for a time from the things which are wont to arouse it.

Anger is like all poisonous kinds of serpents and wild beasts, which while they remain in solitude and in their own lairs are still not harmless; for they cannot really be said to be harmless because they are not actually hurting anyone. For this results in such a case, not from any feeling of goodness, but from the exigencies of solitude. When they have secured an opportunity of hurting anyone, at once they produce the poison that is stored up in them, and show the ferocity of their nature. So in the case of men who are aiming at perfection, it is not enough not to be angry with men. I recollect that when I was dwelling in solitude a feeling of irritation would creep over me against my pen because it was too large or too small; against my penknife when it cut badly or with a blunt edge what I wanted cut; and against a flint if by chance when I was rather late and hurrying to the reading, a spark of fire flashed out. Then I could not get rid of my perturbation of mind except by cursing the senseless matter or, at least, the devil. Therefore for one who is aiming at perfection it is not enough that there should be no men who afford occasion for anger. If the virtue of patience have not been acquired, the feelings of passion which still dwell in his heart can equally well spend themselves on dumb things and paltry objects, and not allow him to gain continuous peace.

III

Of a certain brother who tried to avoid the occasions rather than conquer the spirit of anger.

A certain brother was frequently moved to anger while he dwelt in a monastery. He said, therefore within himself, "I shall go forth into solitude, and when I have no one to quarrel with I shall find rest from this spirit of anger." So he went and dwelt in a certain cave. One day, after he had filled his pitcher and placed it on the ground, it was suddenly upset. Three times he filled it, and three times in the same way the water was spilled. Then, in a rage, he seized the vessel and broke it. When he came to himself, and began to consider how he had been trapped by the demon of anger, he said, "Lo, I am here alone,

84

and yet I have been vanquished by anger. I shall return to my monastery, because, wherever there is most need of striving and of patience, there, no doubt, chiefly is the grace of God to be found." Then, rising up, he returned to his own place.

IV

How by gentleness we may overcome another's anger.

A certain old man had a faithful disciple. Once, in a fit of anger, he drove him from his cell. The disciple waited all night outside the door. In the morning the old man opened it, and, when he saw him, was struck with shame, and said, "You are my father now, because your humility and patience have conquered my sin. Come in again, and from henceforth be you the elder and the father. I will be the disciple, for you have surpassed me, though I am aged."

V

The advice of an elder, showing how we may avoid feeling angry with those who injure us.

A certain monk was injured by one of the brethren. He told what had happened to one of the elders, who said to him, "Let your mind be at peace. The brother has done no injury to you, nor must you think he wished to. He has done injury to your sins. In any trial which comes to you through man, do not blame the man, but just say, `On account of my sins this thing has befallen me.'"

VI

He who is a slave to anger is not likely to conquer other sins.

A certain one said, If a man cannot bridle his tongue in the moment of anger, he will certainly not be able to be victorious over any lust of his flesh.

CHAPTER 14

On Avoiding Many Words

Every idle word that men shall speak, they shall give account thereof in the day of judgment.

- *St. Matt.* xii. 36.

Flee from the throng of the world into the wilderness as much as thou canst; for the talk of worldly affairs is a great hindrance, although spoken of with sincere intention.

Oftentimes I wish that I had held my peace sooner than have spoken; and that I had not been in company.

An evil custom, and neglect of our own good, giveth too much liberty to inconsiderate speech.

- *The Imitation of Christ*, i. 10.

SINS of the tongue are such things as blasphemy, lying, filthy talking, violent and abusive language. These things are admittedly displeasing to God. They are sins, and therefore to be repented of by the man who wishes to enter into life. Outside of these altogether there lies a region of activity for the tongue which can neither be described as good or evil. The common intercourse of daily life, the talks of friends over the fireside, the passing remarks to acquaintances in the streets can neither be branded by the severest rigorist as sinful, nor, except very rarely, are they elevated into the atmosphere of the spiritual by any distinctly religious tone. For most men and women religion has simply nothing to do with their ordinary conversation except in so far as its precepts safeguard the talker from untruthfulness, uncharitableness, and so forth. The hermits took a different view of common talk. To them all talking, even talking about religion, was dangerous. They neither thought nor said that the intercourse of man with man was sinful. When they taught their disciples to learn the habit of silence, it was because silence was safe, not because talking was in itself wrong. It is clear at once that certain kinds of common talk are, as the hermits thought, dangerous. For instance, gossip - the interested discussion of other men's characters and affairs - is dangerous, because it tends to lead to uncharitable thoughts, and sometimes to untruthfulness. It is clear also that the most innocent possible conversation is dangerous at certain times. For instance, it is a dangerous thing to talk in church, because any talk there is likely to lead the mind away from its proper attitude of devotion. Some people, because gossip is dangerous, try to avoid all conversation about their neighbors. Many people, because talking in church is dangerous, avoid it entirely. The hermits simply applied the same reasoning to all conversation on whatever subject, at whatever time. In the first place they recognized that any talk involved the possibility of sin. Bitter experience led them to see that even a conversation on religion generally left a man some word to repent of. To avoid all unnecessary talking was therefore to avoid all unnecessary risk of sinning. Then also they recognized that conversation widened a man's interests in life and in the world. A man with wide knowledge of affairs, and a keen interest in events, is likely to be a good citizen of an earthly state. Precisely so far as his mind is absorbed by the interests of his country, his town, his neighborhood, or his church, so far is it abstracted from the interests of that heavenly city whose builder and maker is God. This strangely ascetic thought finds frequent expression in the teaching of the hermits. They deliberately aimed at being strangers and pilgrims upon earth - men who were merely passing through a foreign country. They wished to concentrate their interests in the land which they called their home, and therefore to abstract them from all the affairs of earth. They knew that every conversation tended to interest them in this world, to make them in heart less of strangers here and more of citizens. Therefore they taught: - Peregrinatio est tacere - our being strangers depends upon our remaining silent.

Once more. The hermits seem to have realized the curious truth that an emotion is weakened when expression is given to it. By our physical nature we are prompted to cry out when we are hurt, because the pain becomes more bearable if our feelings find vent in cries. Analogous to this is the fact that grief is lessened by the telling of it. He suffers less who openly mourns his loss than he who shuts his grief up in his own heart and endures in silence. The same law certainly works out in the sphere of religious emotion. The man who talks about his religious feelings runs a great risk of dissipating them altogether. No man has more need to "guard the fire within" than the preacher whose duty forces him to be for ever giving utterance to the most sacred feelings of his heart. I conceive that this is what the hermits meant by speaking of the mouth as the door of the heart. The mouth is not a door through which any evil enters. The ears are such doors, or the eyes. The mouth is a door only for exit. What was it that they feared to let go out? What was it which someone might steal out of their hearts, as a thief takes the steed from the stable when the door is left open? It can have been nothing else than the force of religious emotion within them. Words conveyed it to listeners, perhaps; they certainly took from the store within. Thus it was that the hermits not only avoided definite sins of the tongue, not only shrank, as many others do, from specially dangerous kinds of talk, but aimed at reducing to the narrow limits of what was absolutely necessary all talk, even talk about religion, and set up as an ideal a life of almost unbroken silence.

I

How the abbot Macarius bid his disciples flee from idle talking.

Once the abbot Macarius, after he had given the benediction to the brethren in the church at Scete, said to them, "Brethren, fly." One of the elders answered him, "How can we fly further than this, seeing we are here in the desert?" Then Macarius placed his finger on his mouth and said, "Fly from this." So saying, he entered his cell and shut the door.

The mouth is the antechamber of the heart.

A certain brother said to the abbot Sisois, "I desire to keep my heart safe from defilement." The old man replied, "It is not possible to guard our hearts while our tongues, by idle talking, open the doors that lead to them."

III

How a man, though he live among friends, may yet be a "stranger and a pilgrim " all his days.

The abbot Sisois said, "Our being strangers and pilgrims consists in this, that we keep continual guard over our tongues."

IV

Silence and solitude are better teachers than much listening to other men and talking to them.

A certain brother asked the abbot Moses to speak some word to him. The old man replied, "Go and sit in your cell. Your cell is well able to instruct you in all things if you remain in it. As a fish that is taken out of the water soon dies, so a monk perishes if he remain long outside his cell."

V

How the value of silence was revealed to Arsenius.

The abbot Arsenius, while he still dwelt in the emperor's palace, prayed to the Lord, saying, Lord, show me the path of salvation." There came a voice to him which said, "Arsenius, fly from the society of men, and thou shalt be saved." When he was on his way to embrace the monastic life he prayed again, saying the same words. He heard then, also, a voice which said to him, "Arsenius, fly, be silent, be in quietness; these are the first steps in learning not to sin.

VI

The abbot Moses praises silent meditation.

A brother in Scete once came to the abbot Moses seeking a word of exhortation. The old man said to him, "Why do you come to me to be taught. Go, sit in your cell. Your cell will teach you all things."

VII

How safety is to be found in silence.

The abbot Nilus said, He remains unhurt by the arrows of the enemy who loves silence, but he who mixes with the multitude gets many wounds.

VIII

How an old man rebuked certain brethren for their many words.

Certain brethren who wished to visit the abbot Antony embarked in a vessel in order to travel to his hermitage. In the same vessel there was an old man who was also going to see St. Antony, but the brethren did not know him. While they sat in the vessel these brethren conversed about an exhortation which they had heard from the fathers, about the Scriptures, and about the work of their hands. All the while that they were talking the old man remained silent. It was not until they came to the place where they disembarked that they knew that he was going to see St. Antony. When they arrived the abbot Antony said to them, "No doubt you found this elder a good companion on your way." Then he said to the old man, "Did you find them good fellow-travelers?" He said, "They are good men enough, but their house has got no door to it. Anyone who wishes can enter into the stable and steal the steed." This he said, because whatever came into their hearts straightway found utterance through their mouths.

IX

Of the extreme difficulty of bridling the tongue.

Once the abbot Sisois said, Truly for thirty years I have not sought God's help against any sins so earnestly as against those of the tongue. Whenever I pray I say this, "Lord Jesus Christ protect me from my tongue." Yet until now I sin through it, and fall through it every day.

X

Of the danger of idle talking.

The abbot Hyperichius said, The serpent whispered to Eve, and she was cast out of Paradise. He who gossips with his neighbor is like unto the serpent. He causes the loss of the soul of him who listens, and his own soul shall not be safe.

XI

How he who is silent, prays.

The abbot Isaiah said, A certain priest was entertaining some of the brethren. While they were eating they talked without ceasing to each other. At last the priest rebuked them, saying, "Be silent, brethren. Lo, there is one among you who does not talk, and his silence ascends like a flame of prayer in the sight of God."

CHAPTER 15

On Evil Thoughts

These are the things which defile a man.

Finally, brethren, whatsoever things are true, whatsoever things are honest, whatsoever things are just, whatsoever things are pure, whatsoever things are lovely, whatsoever things are of good report; if there be any virtue, and if there be any praise, think on these things.

- *Phil.* iv. 8.

O Lord, my God, be not Thou far from me; my God, make haste to help me: for there have risen up against me sundry thoughts, and great fears afflicting my soul.

Do, O Lord, as Thou sayest, and let all evil thoughts fly before Thy face.

- *The imitation of Christ,* iii. 23.

THE necessity for struggling against evil thoughts occupies, as we might expect, an important place in the hermits' scheme of the religious life. The circumstances under which they lived afforded ample opportunities for all kinds of thought and meditation. Often for whole days literally nothing happened to distract the mind from its own musings. The voices of the world were silenced. Only occasionally faint rumours of great events reached the lauras in the desert. The isolation of even those of the Lower Egyptian hermits, who came nearest to living a community life, was for five days of the week almost complete. Other cells were in sight. The figures of other hermits could be descried going for their water-supply or toiling in their gardens. Yet, save for the weekly gatherings on Saturdays and Sundays, there was, under ordinary circumstances, little or no intercourse even between members of the same laura. The rare advent of some stranger might bring the hermits swarming from their cells to bid him welcome; an event of peculiar importance might set the abbot's rude bell ringing to summon the brethren to a consultation; but, as a rule, the life was solitary, and there was little or nothing in its outward circumstances to distract the mind. The work of mat-weaving and basket-making became, for their skilled fingers, purely mechanical. The thoughts were elsewhere even while the hands were busy. So it came that thoughts were not, as they are for men who live amid the world's hurried happenings, swift reflex responses to the excitements of impressions from outside, but wrought out mindpictures and imaginings of things on earth and things in heaven. We think of such day-dreams as the result of the mind's working upon the recollection of experiences long past, or its effort to realize the imagery of Holy Scripture. The hermits conceived them as the result of the mysterious suggestions of powers outside themselves, powers bent upon the conquest of their minds for good or evil. Thus when Isidore showed the abbot Moses the vision of Dothan he displayed a picture of what seemed to him to be literally taking place around the mind of every hermit. The demons never ceased suggesting evil thoughts. The hosts of angels crowded round with thoughts of what was holy and honest, and of good report.

Though the battle was thus being fought by powers outside himself, the hermit was no passive spectator, nor his mind the mere booty of the victorious side. He himself took an active part - indeed, bore the chief share in the strife. On him depended, in the end, the issue of the conflict. It was, indeed, beyond his power to prevent the suggestions of the demons. He could not check the entrance of evil thoughts into his mind. He was, however, able to prevent the evil from obtaining a lodgment in his mind. He could refuse to dream and meditate on thoughts of pride, or hatred, or impurity. According to the vivid imagery of one of their teachers, the mind was a house into which the devil cast sordid things. It was the part of the good householder to pitch them out again speedily, before their accumulation made the home uninhabitable for what was good. Or, as another taught, the evil thoughts might be smothered and packed away, given no opportunity to develop their horrible nature, until, like garments shut unaired into boxes, they moldered into decay.

The advice of the teacher who would have us struggle against only one kind of evil thought, since for each man there is one from which all others draw their power, is suggestive of some deep spiritual experience. It seems as if there is in each soul some one weak point where, once the entrance is won by the demon who assaults it, all other demons are easily able to follow him. Thus to him who has given way to dreams of pride there comes a time when avarice and lust will obtain possession also of his mind. For each man, therefore, it is necessary only in reality to set himself to strive with one kind of evil thought.

While the hermits felt the necessity for watchfulness and struggle, lest they should fall, they gladly recognized that it was through the same strife that they obtained the chance of rising. It is, they taught, through evil thoughts that men make shipwreck of their souls, but also it is through evil thoughts that men are crowned. To them it did not seem a desirable thing to be freed, if that were possible, from the suggestions of evil. What they did wish was to meet the evil at its strongest, and then, through Christ, to vanquish it. To have no evil thoughts is to be no better than a beast. To be afflicted with them, and yet conquer them, is to rise into communion with God.

There are infirmities of the mind, like forgetfulness, which are not evil save in so far as they hinder the soul from the highest flights of all. To those who suffered thus the fathers were very tender. It is most comforting to read the gentle parable by which the brother was encouraged who was unable to bear in mind the religious exhortations which he heard.

In all their teaching about the struggle against evil thoughts the hermits recognised that the truest victory is to be obtained by filling the mind with holy imagery. It is not enough to cast the demons out. We must welcome the angels when they come, must store the mind with good thoughts by constant reading and repetition of Holy Scripture, must keep it stretched in meditation upon the love and the work of the Lord. This, if we can perfectly accomplish it, will certainly give us the victory over evil thoughts, and reduce to impotence the demons who suggest them.

I

Of a certain brother who was continually on the watch against evil thoughts.

It is related that seven brethren used to dwell together on the mountain of St. Antony. At the time of the date-harvests one of them used to be always keeping watch, so as to drive away the birds from the dates. One of the seven, an old man, when it came to his turn to guard the dates, spent the day in crying out, "Depart from within, ye evil thoughts; depart from without, ye birds."

II

The abbot Pastor teaches that evil thoughts are not to be avoided, but overcome.

A certain brother came to the abbot Pastor, and said, "Many evil thoughts come into my mind, and I am in danger through them." The old man led him out into the air, and said to him, "Stretch yourself out, and stop the wind from blowing." The brother, wondering at his words, replied, "I cannot do that." Then the old man said to him, "If you cannot stop the wind from blowing, neither can you prevent evil thoughts from entering your mind. That is beyond your power; but one thing you can do - conquer them."

III

The teaching of the abbot Moses on the same subject.

It is impossible for the mind not to be approached by thoughts, but it is in the power of every earnest man either to admit them or reject them. Their rising does not depend upon ourselves, but their admission or rejection is in our own power. The movement of the mind may well be illustrated by the comparison of a mill-wheel. The headlong rush of water whirls it round, and it can never stop its work so long as it is driven by the water. Yet it is in the power of the man who directs it to decide whether he will have wheat, or barley, or darnel ground by it. For it must certainly crush that which the man in charge of it puts in. So the mind is driven by the torrents of temptation which pour in on it from every side, and cannot be free from the flow of thoughts, but the character of the thoughts we control by the efforts of our own earnestness.

IV

The abbot Pastor speaks of a way in which we may overcome evil thoughts.

The abbot Isaiah once asked the abbot Pastor about evil thoughts which troubled him.

Pastor answered him, "Just as clothes which are put away for a long time in some trunk, and not taken out at all, molder and decay, so the evil thoughts of our hearts, if we do not put them into action, after a long time will fade away.

V

The abbot Moses speaks also of a way of overcoming evil thoughts.

We must constantly fall back upon meditation on the Holy Scriptures, and raise our minds towards the recollection of spiritual things, and the desire of perfection, and the hope of future bliss. In this way spiritual thoughts are sure to arise in us, and our minds will dwell on the things on which we have been meditating. If we are overcome by sloth and carelessness, and spend our time in idle gossip, or if we are entangled in the cares of this world and unnecessary anxieties, the result will be that tares will spring up in our hearts and take possession of them. As our Lord and Savior says, Wherever the treasure of our works or purpose may be, there also our heart is sure to continue.

VI

Of the infirmity of forgetfulness, and how we ought not to despond because of it.

A certain brother said to one of the elders, "Lo, my father, I frequently consult the elders, and they give me advice for the salvation of my soul, yet of all that they say to me I can remember nothing." Now it happened that there were two vessels standing empty beside the old man to whom he spoke. He therefore said to the brother, "Go, take one of the vessels. Put water in it. Wash it, and pour the water out of it again. Then put it back, clean, into its place." The brother did so. Then said the old man, "Bring both vessels here. Look at them carefully, and tell me which is the cleaner." "Surely," said the brother, "that is the cleaner which I washed with the water." Then said the old man to him again, "Even so it is, my son, with the soul which frequently hears the words of God. Even although the memory retain none of them, yet is that soul purer than his who never seeks for spiritual counsel."

VII

Advice for the conquering of evil thoughts.

A certain brother once asked one of the elders, "How shall I overcome the evil thoughts which ceaselessly trouble me?" The elder said to him, "Do not attempt to strive with all of them. Strive only against one. All evil thoughts have a single head and source. - In one man it is this, in another that. It is necessary, first of all, to find out each man for himself what is the origin of his evil thoughts. Then let him bend his energies to the conquest of that one thing, and all other evil thoughts will give way before him."

VIII

That evil thoughts are evil deeds.

"Brethren," said a certain elder, "you are striving to commit no evil deed. I beseech you strive, at the same time, to think no evil thought."

IX

How temptation is not sin, but the means of being good.

A certain elder said, God will not condemn us because evil thoughts enter our hearts, but only if we make a bad use of our evil thoughts. It happens sometimes that men's souls are shipwrecked through evil thoughts, but also it is by the entering in of such thoughts that we become worthy of being crowned.

X

How we are to deal with evil thoughts.

A certain elder said, The devil is an enemy, and your mind is a house. The enemy ceases not to throw into your house every kind of filth that he can find, and to pour into it a world of sordidness. It is your part to be diligent in casting out of your habitation what he throws in. This if you neglect to do, your house will soon be filled with sordid things, and even you yourself will strive in vain to enter into it. Therefore, from the very first, cast out bit by bit everything that he puts in. Then will your house remain clean for you, by the grace of God.

XI

Of our strife against evil thoughts.

A certain elder said, If we have no evil thoughts we are no better than the beasts. The enemy does what is in his power when he suggests them to us. Let us also do the duty which lies within our power. Be instant in prayer, and the enemy will flee. Find time for meditation on divine things, and you will conquer. Persevere, and the good in you will win. Strive hard, and you will be crowned.

XII

How the abbot Moses saw the vision which once the servant of Elisha saw, and was strengthened.

Once, while the abbot Moses dwelt in the region called Petra, he was attacked by the demon of impurity with such fierceness that he could not remain in his cell, nor dared he be alone. He went, therefore, to the holy abbot Isidore and told him of the vehemence of the evil thoughts which came to him. The abbot Isidore bid him be of good cheer, and brought forth from the Holy Scriptures many words of encouragement and strength. Then he bid Moses return to his cell. But this Moses was not willing to do, dreading still to be alone. Then Isidore led him up to the hill which was behind his cell, and said to him, "Turn your eyes westwards and look." He gazed as he was bidden, and beheld a host of demons. Their regiments swept passionately past. They seemed as those prepared for battle, and eager for strife. Then said the abbot Isidore again, "Turn your eyes to the east and look." He gazed as he was bidden, and beheld a numberless array of holy angels. They seemed more glorious and splendid than the shining of the sun, and marched as the army of the good powers of heaven. "Behold," said Isidore, "those whom you saw in the west are the powers which fight against the saints of God. Those whom your eyes looked on in the east are they whom God sends to help His saints. Be sure that the army which fights for us is the stronger one, as saith the prophet Eliseus. Truly, also, St. John saith, 'Greater is He that is in us than he that is in the world.'" When he heard these words Moses took heart of grace, and, being comforted in the Lord, returned to his cell. There he gave God thanks, and praised the long-suffering and the kindness of our Lord Jesus Christ.

XIII

How a certain elder overcame the evil thought which prompted him to postpone his penitence.

It is told of a certain elder that very often his thoughts said to him, "Let to-day go by. Tomorrow will be time enough to repent." He always answered them, "I cannot do this, because to-morrow some other part of God's will must be worked out in me."

CHAPTER 16

On the Life in the World

This day is salvation come to this house, forsomuch as he also is the son of Abraham.

- *St. Luke* xix. 9.

This the Lord said, rebuking those who thought that Zacchaeus was outside the region of the grace of God.

It is not granted to all to forsake all, to renounce the world, and to undertake a life of religious seclusion.

- *The Imitation of Christ*, iii. 10.

THE hermits succeeded in separating their lives not only from the world but from the ways of those Christians who lived in the world. Save for their own brief excursions into village market-places to sell their baskets, and the visits of pilgrims in search of teaching or healing to their cells, the hermits came very little into contact with ordinary members of the Church. It is not to be supposed, therefore, that they either gave much thought to the position of Christians in the world or tried to persuade them to leave it. The hermits were neither theorists nor philosophers. Their religion was entirely practical, and mainly personal. They made no effort whatever to explain why some Christians married, grew rich, and accepted the world's honors, while others retired into the solitude of the wilderness. The hermit was very vividly conscious of his own call to the ascetic life, but he was content to leave others to work out for themselves their own salvation in their own way. The question of the relation of the monastic to the secular life had occupied the mind of Origen, but the hermits either did not know or were totally uninterested in his speculations. The same problem came up for solution afterwards, and was argued out by men like St. Ambrose and St. Augustine, but the hermits did nothing towards providing a philosophy of the life they lived. In spite of the mass of teaching that they left behind them, references of any sort to Christians who lived in the world are extremely few.

The spirit of these few references is wholly different from what we might expect. Experience teaches us that men who are rigorists, who, to a greater or less extent, stand aloof from the common joys and labors and ambitions of mankind find it necessary, as it were in self-defense, to judge sternly of those who do not walk in their ways. It is a lamentable fact that the great earnestness which enables men to make real renunciations is too often connected not only with want of charity, but with a total incapacity to appreciate the amount of genuine religion which exists in systems less rigorous than their own. It has come to be recognized as almost an unvarying law that the Christian who fasts and weeps, even if he does not fail in charity to individuals, will never be able to recognize that there is a real religion in which laughter and dancing find their place. Of all men the hermits were the most rigorous in their life. We should expect therefore to find them most ready in definite condemnation of religious ways which differed from their own. I do not suppose that anyone who has learnt to appreciate the depth and spirituality of their religion would expect to find them bitter and uncharitable towards individuals. Such a spirit cannot coexist with the seeing and desiring to see the God who is love. Nor, I think, should we be surprised to find them recognizing some possibility of good in the life of the Christian in the world. It is, however, with real amazement that we read the few judgments which they passed on the secular life. It is not that they look on such life as good, though poorer and lower than their own; still less do they regard it with that pitying contempt which is often misnamed charity. They recognize gladly that it may be in every way equal to their own lives. They go back to their cells from the kitchens of housewives and the workshops of tradesmen humbled by the contemplation of a perfection to which they themselves have not been able as yet to attain.

St. Macarius of Alexandria was one of the very sternest of the hermits in his ascetic practices. The fierceness of his efforts to subdue his body shock us, while we wonder at the strength of the man who made them. Of all the leaders of the movement he would seem the least likely to appreciate the beauty of a Christian life lived in the world. Yet it is he who says, "Truly virginity is nothing, nor marriage, nor the monk's life, nor life in the world." Certainly it was a special revelation which led him to the house of the two women whose way of life taught him this truth; yet we must suppose an almost incredible magnanimity in the man, placed as St. Macarius was, who could receive and profit by such a revelation. It is not so wonderful that St. Antony should have reached to the understanding of the many different ways in which God leads men upwards to Himself. We know enough about him to appreciate the broadness and sanity of his character. Yet even from him it is startling to hear such words as those he spoke to the Alexandrian tanner: "Of a truth, my son, you are on your way to the kingdom of God, and I, like a man without wisdom, am passing the time of my solitude without attaining to the measure of the perfection that you have told me of."

The words of Muthues are poorer, perhaps, than the confessions of St. Antony and St. Macarius, yet they have a special value. They show us how it was that the hermits became capable of such clear-sightedness in the recognition of good. It was through their humility, that virtue which is likened, aptly, to the rudder of a ship. God Himself could not have revealed the great truths about life, which these saints saw, except to men whose hearts were well prepared for His Spirit by a long discipline of subduing pride.

I

How the divine guidance enabled St. Antony to see that a life well pleasing to God may be accomplished by one who is in the world as well as by a monk.

Once, while St. Antony was praying in his cell, there came to him a voice which said, "Oh, Antony, for all your life in the desert you have not yet attained the measure of the perfection of a tanner who lives in Alexandria." When he heard this the saint rose up early, took his staff; and came with haste to Alexandria. He speedily found the man of whom he had been told. The tanner was struck dumb at the sight of so great a saint. St.

Antony said to him, "Describe to me the manner of your life. I have come here from the desert to learn about your good deeds." The tanner answered him, "I have not, so far as I know, done anything good at all. I am a very sinful man. When I rise from my bed in the morning, before my work begins I say, 'All the people in this city must be better than I am. From the least to the greatest they may well be entering into the kingdom of heaven. I, because of my sins, am certainly going to everlasting punishment.' Then when I am going to rest at night I find myself obliged to repeat this same saying." Then St. Antony replied to him, "Of a truth, my son, you, as you sit here quietly in your house, are on your way to the kingdom of God. I, like a man without wisdom, am passing the time of my solitude without attaining to the measure of the perfection that you have told me of."

II

How St. Macarius was guided by the Spirit to a knowledge of the same truth.

Once, while the abbot Macarius was praying, a voice sounded in his ears, which said to him, "Macarius, you have not yet arrived at the measure of the sanctity of two women who dwell in the neighboring city." When he heard this he arose and, taking his staff, set forth for the city which had been named. He sought and found the house where the women lived. When he knocked at the door one of the women came out, and, perceiving who he was, welcomed him into the house with great joy. St. Macarius called the two together to him, and said, "On your account I have endured the toil of coming here from my solitude. I desire to know your way of life. I pray you to describe it to me." They, however, replied to him, "Most holy father, what kind of life is ours for you to ask about?" He persisted in asking that they would describe it to him. Then, since he compelled them, they said, "We are not, indeed, related to each other by blood, but it happened that we married two brothers. Now, though we have lived together for fifteen years, we have had no quarrel, neither has either of us spoken a sharp word to the other. We both desired to leave our husbands and enter a community of holy women. We begged our husbands to permit us, but they would not. Then we vowed that until the day of our death we should hold no worldly talk with each other, but converse only about spiritual things." When St. Macarius heard what they told him, he said, "Truly virginity is nothing, nor marriage, nor the monk's life, nor dwelling in the world. It is purposes and vows like this which God seeks from us, and He gives the spirit of life to all alike."

III

How the monk must not reckon himself safe because he is a monk, nor must think of those who live in the world as lost.

The abbot Muthues said, The nearer a man draws to God the more he sees his own sinfulness. Thus when the prophet Isaiah had his vision of God he exclaimed that he was wretched and unclean. Let us be careful to hold this truth fast, for the Scripture saith, "Let him who thinketh he standeth take heed lest he fall." We voyage doubtfully across the waves of this world. We indeed may seem to be sailing over quiet seas while they who dwell in the world go amid dangers. We shape our course in the daylight, lit upon our way by the Sun of Righteousness. They, as if in the night-time, may steer in ignorance of where they go. Yet it often may come to pass that the dweller in the world, just because he voyages through a dark night, is very watchful, and his ship comes safe to port. So too we, just because we voyage over quiet seas, grow careless. Too often from our very security we perish, letting go the helm, which is humility. Just as no ship can be safe without a rudder, so it is impossible for a man to come safe to his journey's end without humility.

CHAPTER 17

The Inner Life and the Visible Church

The Scribes and the Pharisees sit in Moses' seat: all therefore whatsoever they bid you observe, that observe and do; but do not ye after their works.

- *St. Matt.* xxiii. 2, 3.

DURING the earlier stages of the monastic movement the hermits came very little into contact with Church authority. They lived, at first, outside the sphere of clerical activity. They were often far out of reach of village churches, and a priest in order to minister to them must have been himself prepared to become a hermit. They were, I believe, at first almost entirely uninterested in the controversies which rent the Church. Their devoted loyalty to St. Athanasius was less the result of their dogmatic orthodoxy than a tribute to the noble unworldliness of the great patriarch's character. For them the entire interest of religion centered in the effort to keep the commandments of God and follow the example of Jesus Christ. Afterwards, of course, their spirit changed, and they became the earnest and sometimes even fanatical opponents of positions deemed heretical. Long before that time came, however, they had been obliged to adjust their relations to ecclesiastical authority.

It is not to be supposed that even the earliest hermits were in any way hostile to the clergy or opposed to the system of Church government, still less that they were contemptuous of the means of grace committed to the Church's guardianship. Rather, we must think of them as men so absorbed in fostering and perfecting the inner life of personal communion with God, that they did not feel the need of absolution or of Sacraments. It was inevitable that as their numbers grew, and as they gathered into the communities of the lauras, this position must give way. The change was a very critical one. There was the possibility of a revolt against all the external machinery of religion. It is quite easy to understand that this was the most likely consequence of the earlier aloofness. Men who are genuinely on fire with a love for holiness are sure to resent the marks of corruption and insincerity which must ever be visible in the garments of the Church on earth. Men of intense spirituality are likely to revolt against the claims of authority which sometimes must seem to break in upon their own communion with God. It is not the least wonderful thing in the history of Egyptian monasticism that it never produced even the beginnings of a schism.

The change from the original position of entire spiritual independence to that of faithful loyalty to the Egyptian patriarchs took place silently, and has left but few traces of the steps by which it was accomplished. The two stories which form this chapter are quoted as examples of the way in which the hermits learned their lesson of obedience. They furnish us, I think, with valuable spiritual lessons, and give evidence of a grace in their heroes which is very worthy of imitation.

I

Of a hermit who refused the ministration of a priest who was a sinner.

Once a man said to a certain hermit, "The priest who ministers to you is a sinner." Then doubt concerning the priest took possession of the hermit's mind, and when, according to his custom, the priest came again he shut the door against him. There came a voice to him as he sat in his cell, which said "Assuredly men are governed by someone else than me." Then he beheld a vision. He stood in a great garden wherein were fruit trees of every kind. He saw there the engine by which water was raised from the river for the watering of the garden, and lo, all the vessels connected with it were of gold. He was about to drink of the water when he saw that the man who tended the engine was a leper, loathsome to behold. Then all desire of drinking departed from him. There came the voice and spoke to him again, "Oh man, have you beheld the beauty of the garden and the trees? Have you seen the wheel with its golden furniture? Have you seen, too, the gardener and the misfortune which has overwhelmed him ?" The hermit answered, "I have seen all this." Then said the voice, "Does his disease injure at all the trees or the beauty of the garden?" And he answered "No." The voice said to him, "It is even so with the priest who makes the sacrifice. He may be a sinner, but his sin diminishes nothing of the honor due to the body of the Lord. The divine virtue is ever active in the Eucharist. The prayers with which he celebrates are always the same as the prayers of holy fathers."

II

How the Lord himself taught the abbot Schnoudi the respect due to those who sit in Moses' seat.

It happened one day that the abbot Schnoudi was holding converse with our Savior Jesus Christ, when the Bishop of Schmin arrived at the monastery. He sent to ask the abbot to come to him that he might talk to him. But Schnoudi, the Savior as has been told being with him, sent back a message by the servant, "Schnoudi at this time has no leisure." When the servant had given this message to the bishop he sent again, saying, "Bid him be kind to me, for I have come here for the purpose of knowing him." But Schnoudi said to the brother who brought the message, "Tell him again that I have no leisure to see him." Then the bishop was vexed, and said, "Say to him, If you do not come I shall excommunicate you." Schnoudi, when he heard the message, smiled and said, "Behold the folly of this man of flesh and blood. Lo, here is with me the creator of heaven and earth. I shall continue to abide with Him." Then the Saviour Himself spoke, and said to him, "Oh, Schnoudi, rise and go to the bishop, lest he excommunicate you. If he does I shall not receive you into heaven. The Father promised, saying, 'Whatsoever thou shalt bind on earth shall be bound in heaven, and whatsoever thou shalt loose on earth shall be loosed in heaven.'" Then when Schuoudi had heard these words he hastened to the presence of the bishop.

Chapter 18

In the Hour of Death

If a man keep My saying, he shall never see death.

- *St. John* viii. 51.

Love Him and keep Him for thy friend, who when all go away will not forsake thee, nor suffer thee to perish at the last.

- *The Imitation of Christ*, ii. 7.

In the hour of death, and in the day of judgment, Good Lord, deliver us.

- *The Litany.*

THE tendency which sometimes manifests itself among pious people to think much of the last moments of those who depart hence in the faith of Christ is certainly morbid and leads to devotional thought of a sentimental kind. For most men the arrival of the supreme moment has been preceded by a weary period of physical suffering. The body is worn and wasted. The mind has lost its power, even the power of expectation. The spirit is depressed with weakness, sympathetic with the decay of the home God formed for it. Neither brave words nor clear vision of what lies beyond are commonly to be looked for. Very often, too, a certain self-forgetfulness which we cannot but praise fixes the thought of him who is to go more on the grief of the parting and the foreseen desolation of those who are left, than on the hope of the breaking forth of glory when the veil is lifted. Then from the mouth of the dying believer come, haltingly, words meant to be comfortable. They form no message from beyond. That they are spoken from the grave's brink adds only a great pathos to the familiar attempts at consolation with which we who are left try to lighten the mourners' grief. Yet sometimes, even in the hour of death, the spirit is so far triumphant over bodily decay as to recognise the supreme importance of the crisis through which it passes. The world beyond is realized, is felt - we may say, is seen. This world, and all that life in it has meant, is seen too, not any longer as a succession of incidents of which the nearest alone seem great, but seen whole with all its days in true perspective. From such vision there is every hope that we should learn. The Christian will not indeed expect or hope to grasp at secrets unrevealed, but he may reverently expect to be told of what the great emotion will consist. Is it to be joy or fear? Shall we be absorbed with regret looking backwards, or rapt in expectation of what is to come? Will joy and fear, regret and hope, all alike yield to an overmastering curiosity about what that other world is like? Will doubt for the last time harass us, or may we look for the extreme beatitude of the satisfaction at length of desire for the Beloved?

Of the four death-bed stories which I tell, one seems at first to speak only of regret for past mistakes. The Archbishop Theophilus tells us only that he knew at last that death ought to have been more often present in his mind. No doubt he was conscious that he would have lived better had he lived more as one who was about to go. Yet even here it is possible to feel that he regretted not only a mistake, but the missing of a great source of joy. If the passing away was a glad thing to him when it came, he would regret that he had failed to get the joy of its anticipation. The abbot Pammon saw his past life in the light of that which was dawning for him. We catch in his summary of his life's accomplishments something of the triumph of St. Paul's - "I have fought the good fight. I have finished my course." Yet all that he was, or did, or felt seems nothing to him in comparison to the vista of devotion which stretches before him. Some regret there is in what he says, but in the main he speaks to us of expectation. To the abbot Agathon the hour of death brings a certain doubt. He, too, sees the life that is past, but his vision of that which is to come stops short at the judgment act. God is to pronounce that he has done well or been mistaken. He is not sure, even at the last, what God's pronouncement is to be. This is his doubt. But it is a doubt which neither terrifies nor unmans, for it is covered by a larger faith. The pronouncement is to be God's. That insures that it, at least, will be just and right. The man may have been mistaken. Death takes him where his mistake is surely to be rectified. For Sisois the hour of death brings an unspeakable rapture. St. Antony is with him, and the prophets and the apostles. He speaks to the angels, and they to him. Death means union with all whom he loved best. It is the satisfaction of long unfulfilled desire. Yet even for him there is regret. He knows that he is not good enough to join such company. Sins only half repented of crowd to his remembrance. He asks for time for more repentance. He is answered by the beatific vision of the Lord Himself, and love made perfect casts out fear.

I

How in the hour of death Pammon was fain to confess that his service of God had been but very imperfect.

The abbot Pammon in that hour when he was passing away from the body spoke thus to the other holy men who stood around him, "Brethren, since the day that I came here to the desert, amid built this cell of mine, I do not think that I have ever eaten anything except what the work of my hamids earned. I do not remember that I have reason to repent of any exhortation which I ever gave to the brethren. Yet, if indeed I am now going to God, it seems to me that I have not yet begun to learn to worship Him."

II

How in the hour of death the abbot Agathon, though he knew nothing against himself, yet was not thereby justified.

At the time when the abbot Agathon lay dying his eyes were fixed for three whole days, as if he were in a trance. The brethren who were with him touched him to awaken him, and said, "Father, where are you now?" He replied, "I stand gazing at the God who judges me." Then the brethren said, "Surely you are not afraid." He answered them, "While I was with you on earth, as far as in me lay, I strove to obey the commandments of God. Yet I am but a man, and now I am not sure - how can I be sure? - that the things I did were really pleasing in God's sight." The brethren said, "have you no confidence that your deeds were in accordance with the will of God?" He replied, "I have no confidence now that I am standing in the sight of God. Man judges about what is right and wrong. That is one judgment. God also judges what is right and wrong. His judgment is another and different."

III

The glorious vision of the abbot Sisois in the hour of death.

Many elders gathered round the abbot Sisois when the time of his falling asleep came to him. They saw his face shining with a wondrous radiance, and he said to them, "Lo, the abbot Antony is coming to me." After a little while he said, "The company of the prophets is along with him." Then his face shone with a brighter light, amid he said, "The blessed apostles are beside me." It seemed, then, to those who stood by as if he spoke to someone, and they asked him to tell them with whom he talked. He said, "The angels have come to bear away my soul, and I am asking them to grant me yet a little while for penitence." Then the fathers said to him, "Surely you have no need of penitence?" But he replied, "Verily I say to you that I have never yet grasped even the beginning of true penitence." Then they felt that in him the fear of God was indeed perfected. Suddenly his face was lighted with all the splendor of the sun, and he cried out to them, "Behold, behold my brethren, the Lord Himself is come to me." Then while he spoke these words, his spirit fled, and all the place was filled with a sweet smell.

IV

The words of Theophilus the Archbishop, which he spoke in the hour of death.

Theophilus the Archbishop, of blessed memory, when he was about to depart, said, "Blessed art thou, Arsenius, for thou hast always had this hour before thine eyes."

www.ingramcontent.com/pod-product-compliance
Lightning Source LLC
Chambersburg PA
CBHW051415200326
41520CB00023B/7247